Alfalfa Management Guide

Authors

Dan Undersander
Extension agronomist, forages
University of Wisconsin

Mark Renz
Extension agronomist, weed control
University of Wisconsin

Craig Sheaffer
Research agronomist
University of Minnesota

Glen Shewmaker
Extension agronomist
University of Idaho

Mark Sulc
Extension agronomist
The Ohio State University

Acknowledgments

The authors wish to thank reviewers from industry and various universities for their suggestions and everyone who supplied photos, including those not specifically mentioned:

Steve Bicen, University of Wisconsin *anthracnose; aphanomyces, roots; Fusarium wilt, roots; Phytophthora, roots; root assessment; verticillium wilt, root*

Dennis Cosgrove, University of Wisconsin *autotoxicity*

Jim Ducy *title page photo*

Del Gates, Kansas State University *alfalfa weevils*

Craig Grau, University of Wisconsin *aphanomyces, stunting; bacterial wilt, stunting; black stem, lesions; Fusarium wilt, field; Phytophthora, plant; sclerotinia; stand assessment; verticillium wilt, plants*

B. Wolfgang Hoffmann, University of Wisconsin *alfalfa plant, page 1; alfalfa flowers*

Eric Holub, University of Wisconsin *aphanomyces, seedling*

Jeffrey S. Jacobsen, Montana State University *nutrient deficiencies—all except boron leaf (from* Diagnosis of Nutrient Deficiencies in Alfalfa and Wheat*)*

Pioneer Hi-Bred International, Inc. *alfalfa closeups; cover photo ; cow; inside cover*

Lanie Rhodes, Ohio State University *black stem, leaves; common leaf spot; lepto leaf spot*

Marlin E. Rice, Iowa State University *alfalfa weevil, blister beetles; clover leaf weevils; grasshopper; pea aphids; plant bug, adults; potato leafhopper, adult; spittlebug; variegated cutworm*

Judy A. Thies, USDA-ARS *root-lesion nematodes*

John Wedberg, University of Wisconsin *alfalfa blotch leafminer; clover root curculio, damage*

Thanks also to Bruce Gossen and R'eal Michaud, research scientists at Agriculture and Agri-Food Canada, for their contributions to the disease maps.

This publication is a joint effort of:

University of Wisconsin-Extension, Cooperative Extension

Minnesota Extension Service, University of Minnesota

Iowa State University Cooperative Extension Service

Published by:
American Society of Agronomy, Inc.
Crop Science Society of America, Inc.
Soil Science Society of America, Inc.

© 2015 by American Society of Agronomy, Inc., Crop Science Society of America, Inc., and Soil Science Society of America, Inc.

ISBN: 978-0-89118-347-1

Library of Congress Control Number: 2015956808

Editor: Lisa Al-Amoodi
Designer: Patricia Scullion

Editor previous editions: Linda Deith
Designer previous editions: Susan Anderson

American Society of Agronomy
Crop Science Society of America
Soil Science Society of America
5585 Guilford Road
Madison, WI 53711-5801 USA
TEL: 608-273-8080
FAX: 608-273-2021
www.agronomy.org
www.crops.org
www.soils.org

Contents

Profitable forage production depends on high yields. Land, machinery, and most other operating costs stay the same whether harvesting 3 tons per acre or 6 tons per acre. Top yields in the northern United States have approached 10 tons per acre while average yields are around 3 tons per acre. This booklet describes what it takes to move from a 3-ton yield to 6 or 9 tons per acre.

Establishment

A vigorously growing, dense stand of alfalfa forms the basis for profitable forage production. Profitable stands are the result of carefully selecting fields with well-drained soil, adding lime and nutrients if needed, selecting a good variety, and using appropriate planting practices to ensure germination and establishment.

Select a field carefully

Soil type, drainage, and slope

Alfalfa requires a well-drained soil for optimum production. Wet soils create conditions suitable for diseases that may kill seedlings, reduce forage yield, and kill established plants. You can reduce some disease problems associated with poor drainage by selecting varieties with high levels of resistance and by using fungicides for establishment. Poor soil drainage also reduces the movement of soil oxygen to roots. Poor surface drainage can cause soil crusting and ponding which may cause poor soil aeration, micronutrient toxicity, or ice damage over winter. Even sloping fields may have low spots where water stands, making it difficult to maintain alfalfa stands.

Soils should be deep enough to have adequate water-holding capacity. Alfalfa has a long taproot that penetrates more deeply into the soil than crops such as corn or wheat which have more fibrous, shallow roots. Under favorable conditions, alfalfa roots may penetrate over 20 feet deep. This great rooting depth gives alfalfa excellent drought tolerance.

Sloping fields where erosion is a problem may require use of erosion control practices such as planting with a companion crop or using reduced tillage to keep soil and seed in place until seedlings are well rooted.

Control perennial weeds

Fields should be free of perennial weeds such as quackgrass. If not controlled before seeding, these weeds may re-establish faster than the new alfalfa seedlings and reduce stand density. Weed management is discussed in more detail in the Production section.

Fields should be free from herbicide carryover that may affect growth of the new alfalfa and/or companion crop. This is especially important after drought years and on fields where high herbicide application rates or late-season applications of long-lasting herbicides were used.

Autotoxicity

Alfalfa plants produce toxins that can reduce germination and growth of new alfalfa seedings. This phenomenon is known as autotoxicity. The extent of the toxin's influence increases with the age and density of the previous alfalfa stand when it was killed.

The autotoxic compounds are water soluble and are concentrated mainly in the leaves. The compounds impair development of the seedling taproot by causing the root tips to swell and by reducing the number of root hairs (Figure 1). This limits the ability of the seedling to take up water and nutrients and increases the plant's susceptibility to other stress factors.

Surviving plants will be stunted and continue to yield less in subsequent years (Figure 2). A waiting period after destroying the old stand is necessary to allow the toxic compounds to degrade or move out of the root zone of the new seedlings. Weather conditions influence

Figure 1. Effect of autotoxicity on root development of alfalfa.

To ensure good stands, calibrate seeding depths and rates carefully and plant in a firm, moist soil.

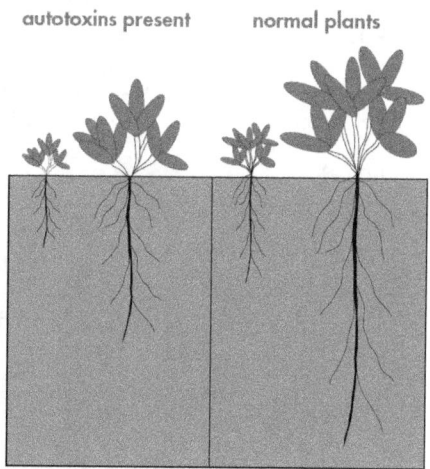

autotoxins present   normal plants

Source: Jennings, Nelson, and Coutts, Universities of Arkansas and Missouri, 1998

the speed with which the toxins are removed. Breakdown is more rapid under warm, moist soil conditions. The autotoxic compounds are removed more rapidly from sandy than more heavy textured soils. However, while the compounds are present, the effect on root growth is much more severe in sandy soils.

Ideally, grow different crop for one season on sandy soils or two seasons on medium to heavy soils after plowing down or chemically killing a 2-year or older stand before seeding alfalfa again in the same field. This is the best and safest way to manage new seedings of alfalfa.

Test soil before planting

Proper fertility management, including an adequate liming program, is the key to optimum economic yields. Proper fertilization of alfalfa allows for good stand establishment and promotes early growth, increases yield and quality, and improves winterhardiness and stand persistence. Adequate fertility also improves alfalfa's ability to compete with weeds and strengthens disease and insect resistance.

Figure 2. Effect of waiting periods when establishing alfalfa following alfalfa. Note that even though the number of plants is similar for all but fields planted with no waiting, yield increases dramatically.

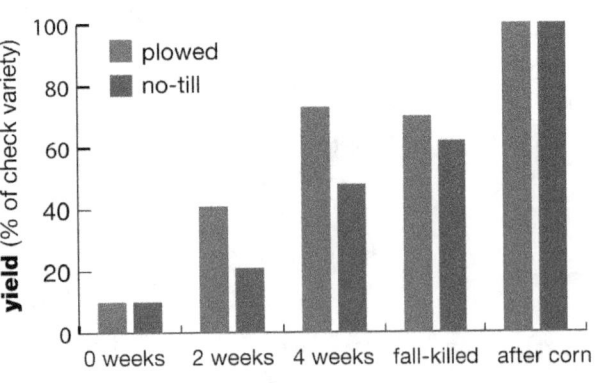

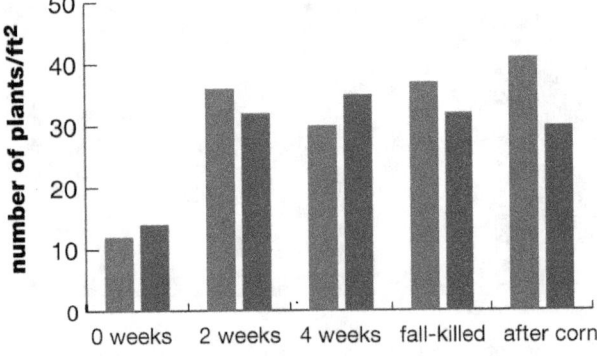

Source: Cosgrove et al., University of Wisconsin-River Falls, 1996.

Effect of autotoxicity on alfalfa stands when alfalfa is seeded (a) immediately following alfalfa plowdown, (b) 4 weeks later, and (c) after 1 year.

Advanced Techniques

Reseeding recommendations

Reseeding stands within 1 year carries a certain amount of risk of yield and stand failure due to autotoxicity. Use Table 1 to calculate the risk. If reseeding, consider the following:

- Never interseed to thicken a stand that is 2 years old or older. Young plants that have been interseeded often look good early but die out over summer because of competition for light and moisture from the established plants.

- In stands where the likelihood of successfully interseeding alfalfa is low, consider interseeding red clover or a grass species such as ryegrass or orchardgrass. These species will most likely establish well and provide good yield until new stands can be re-established.

Reseeding establishment failures

- Disk down a seeding failure and reseed either in the late summer after a spring seeding or the following spring. Autotoxic compounds are not present the first year.

- Reseed gaps in new seedings as soon as possible within 1 year of initial seeding.

Fields differ in their fertilizer needs. Soil testing is the most convenient and economical method of evaluating the fertility levels of a soil and accurately assessing nutrient requirements.

Most soil testing programs make recommendations for pH and lime, phosphorus, and potassium. Soil tests for secondary nutrients and micronutrients are questionable. Optimal soil test levels for alfalfa differ among states due to varying subsoil fertility, nutrient buffering capacities, soil yield potentials, and different management assumptions. For more detailed information on soil test recommendations, contact your local Extension office.

Table 1. Alfalfa autotoxicity reseeding risk assessment.

	points	score
1. Amount of previous alfalfa topgrowth incorporated or left on soil surface		
Fall cut or grazed	1	
0 to 1 ton topgrowth	3	
More than 1 ton topgrowth	5	
2. Disease resistance of the variety to be seeded		
High disease resistance	1	
Moderate disease resistance	2	
Low disease resistance	3	
3. Irrigation or rainfall potential prior to reseeding		
High (greater than 2 inches)	1	
Medium (1 to 2 inches)	2	
Low (less than 1 inch)	3	
4. Soil type		
Sandy	1	
Loamy	2	
Clayey	3	
5. Tillage prior to reseeding		
Moldboard plow	1	
Chisel plow	2	
No-till	3	
6. Sum of points from Questions 1–5		
7. Age of previous alfalfa stand		
Less than 1 year	0	
1 to 2 years	0.5	
More than 2 years	1	
8. Reseeding delay after alfalfa kill/plowdown		
12 months or more	0	
6 months	1	
2 to 4 weeks	2	
Less than 2 weeks	3	
Your total score (multiply points from 6, 7, and 8)		

Alfalfa reseeding risk

If you score:	The autotoxity risk is:	Recommendation
0	low	Seed
4–8	moderate	Caution—potential yield loss
9–12	high	Warning—yield loss likely
>13	very high	Avoid reseeding—likely stand and yield loss

Source: Craig Sheaffer, Dan Undersander, and Paul Peterson, Universities of Minnesota and Wisconsin, 2004.

Apply lime before seeding

Liming is the single most important fertility concern for establishing and maintaining high yielding, high quality alfalfa stands. Benefits of liming alfalfa include:

- increased stand establishment and persistence,
- more activity of nitrogen-fixing *Rhizobium* bacteria,
- added calcium and magnesium,
- improved soil structure and tilth,
- increased availability of phosphorus and molybdenum (Figure 3), and
- decreased manganese, iron, and aluminum toxicity (Figure 3).

For maximum returns, lime fields to pH 6.7 to 6.9. Field trials performed in southwestern Wisconsin show that yields drop sharply when soil pH falls below 6.7 (Figure 4).

Because lime reacts very slowly with soil acids, it should be applied 12 months before seeding. For typical crop rotations, the best time to apply the recommended amount of lime is one year prior to seeding the alfalfa. This allows time for reaction with the soil. In addition, the accompanying tillage for rotation crops may result in two or three remixings of the lime with the soil. By the time alfalfa is replanted, the pH should be raised to the desired level.

Aglime should be broadcast on the surface of the soil, disked in, and then plowed under for maximum distribution and neutralization of acids in the entire plow layer. Plowing without disking may deposit the lime in a layer at the plow sole. For high rates of lime (more than 6 tons/acre), apply half before working the field; work

the remaining half into the soil after plowing or other field preparation.

Lime effectiveness is determined by its chemical purity and the fineness to which it is ground. Figure 5 illustrates the greater effectiveness of more finely ground lime. To achieve the same pH change, coarse aglime must be applied further in advance and at higher rates than fine aglime but is usually less expensive per ton. It may not be necessary to re-lime as often where some coarse lime is used.

When comparing prices, be sure to evaluate materials on the basis of amounts of lime needed to achieve similar effectiveness. The relative effectiveness of various liming materials is given by its lime grade, effective calcium carbonate equivalency (ECCE), effective neutralizing power (ENP), or total neutralizing power (TNP).

Figure 3. Available nutrients in relation to pH.

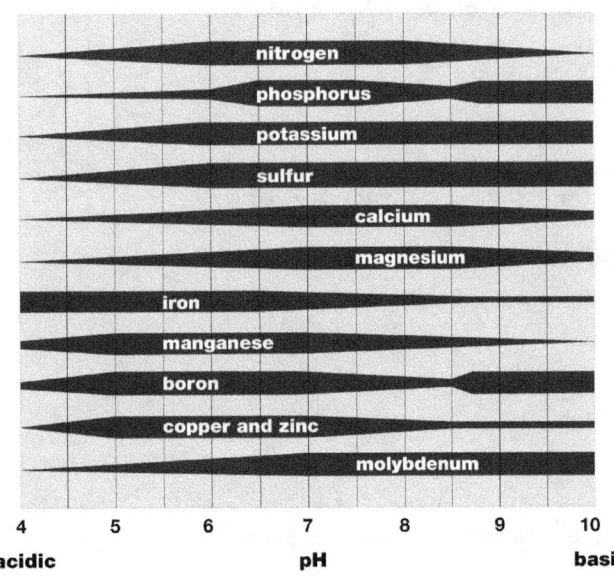

Figure 4. First-cutting alfalfa yield relative to soil pH.

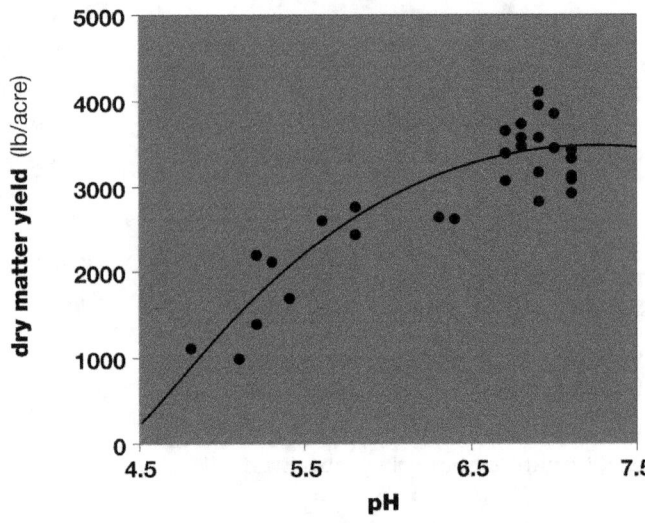

Source: Wollenhaupt and Undersander, University of Wisconsin, 1991.

Liming materials come in several forms. Calcitic products contain calcium-based neutralizers while dolomitic sources contain both calcium and magnesium. Both are effective for changing soil pH. Some claims are made that when the calcium to magnesium ratios in the soil are low, calcitic limestone should be used. Research evidence does not support these claims, as virtually all midwestern and northeastern soils have ratios within the optimal range. Dolomitic limestone itself has a calcium to magnesium ratio within the normal range for plant growth. The addition of calcitic limestone or gypsum for the purpose of adding calcium or changing the calcium to magnesium ratio is neither recommended nor cost effective.

Several by-products, such as papermill lime sludge and water treatment plant sludge, may be used as liming materials. Since the relative effectiveness of some of these materials is highly variable, be sure you know its effective neutralizing power.

Nutrient needs during establishment

Tillage during establishment provides the last opportunity to incorporate relatively immobile nutrients during the life of the stand. Typical nutrient additions tend to include phosphorus, potassium, and sulfur.

Phosphorus. Adequate soil phosphorus levels increase seeding success by encouraging root growth. Phosphorus is very immobile in most soils. Wisconsin research confirms that at low to medium soil test levels, incorporated phosphorus is more than twice as efficient as topdressed phosphorus.

Potassium. Research has shown that although potassium has relatively little influence on improving stand establishment, yield and stand survival are highly dependent on an adequate potassium supply. When soil tests are in the medium range or below, sufficient potassium should be added to meet the needs of the seeding year crop including the companion crop.

Sulfur. Despite the higher cost, sulfur, where needed, is typically topdressed annually with fertilizer rather than incorporated at planting due to ease of application. Elemental sulfur can be used as the sulfur source and may be applied at seeding. Elemental sulfur must be converted to sulfate-sulfur before it can be used by plants. This process is relatively slow, especially when sulfur is topdressed. Therefore, incorporating moderately high rates (50 lb/acre sulfur) of elemental sulfur at establishment will usually satisfy alfalfa sulfur requirements for the life of the stand. Tissue test alfalfa by sampling the top 6 inches immediately prior to harvest. This is the best procedure for determining deficiencies of sulfur and other secondary nutrients and micronutrients. Adequacy levels are shown in the Production section.

Nitrogen. Research has shown that small additions of nitrogen may enhance establishment and seeding year yields. Apply 25 to 30 lb/acre nitrogen when alfalfa is direct seeded on coarse-textured soils with low organic matter contents (less than 2%). Apply 20 to 35 lb/acre nitrogen when seeding alfalfa with a companion crop and apply 40 to 55 lb/acre nitrogen if you will be harvesting the companion crop as silage.

Manure. Manure is a source of macronutrients and micronutrients and can be used to help meet the nutrient needs of alfalfa. Manure testing is recommended prior to application to any cropland.

For application before seeding, manure should be thoroughly mixed with the soil and limited to rates of not more than 7 tons/acre of solid dairy manure or 20,000 gal/acre of liquid dairy manure (environmental requirements may lower the recommended rates).

Figure 5. Lime availability at different particle sizes.

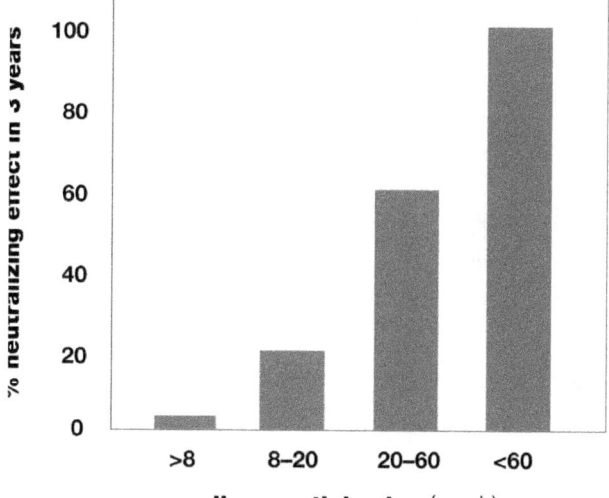

% neutralizing effect in 3 years

aglime particle size (mesh)

Select a good variety

Plant breeders have developed alfalfa varieties with greater yield potential, better disease resistance, and improved forage quality. But with over 250 varieties available, how does one decide? The major factors leading to profitability are:

- yield potential,
- persistence (percent stand remaining or estimated from winterhardiness and disease resistance ratings),
- winterhardiness,
- disease resistance, and
- forage quality.

As illustrated in Table 2, yield has the largest effect on profitability, persistence next, and other factors have a lesser effect. Other factors such as fall dormancy and intended use may be important in certain circumstances.

Yield potential

Look for varieties with high yields in university trials. Compare new varieties against one you have grown. Comparing varieties to the same check, planted within the trial, also allows comparison across several trials. New varieties should perform better and result in higher yields. In Wisconsin and Minnesota variety trials performed between 1980 and 1998, the top varieties have yielded slightly over 1 ton more per acre than Vernal (a standard check variety) for each year of stand life (Figure 6). For short-term stands, select varieties by yield from 2- to 3-year-old stands. For long-term stands, select by yield from 4- to 5-year-old stands.

Varieties will perform differently in various growing regions. Look for top yields of a variety grown in a site with as similar a soil type and climate to your farm as possible. Also, look for top yield over several sites. This indicates stability for high yield and is important because soils may vary on your farm and weather conditions change from year to year.

Persistence

Compare stand survival ratings or yields of 4- to 5-year-old stands to determine relative persistence of varieties. Persistence in northern locations depends primarily on winterhardiness because of the severity of winter temperatures; farther south persistence is more dependent on disease resistance. If stand survival ratings or yields of 4- to 5-year-old stands are not available, use winterhardiness and disease resistance to estimate persistence.

When evaluating varieties, remember that long-term stands are not necessarily the most profitable. Many farmers are finding that a 4-year rotation with 3 years of alfalfa may be more profitable than trying to keep one stand of alfalfa for 5 or 6 years in a 7- or 8-year rotation. This occurs for the following reasons:

- younger stands of alfalfa yield more than older stands,
- with a 4-year rotation, nitrogen credits from plowdown alfalfa are available twice in 8 years,

Figure 6. Yield difference between top and bottom alfalfa entries in Wisconsin Alfalfa Trials, 1985 to 2014.

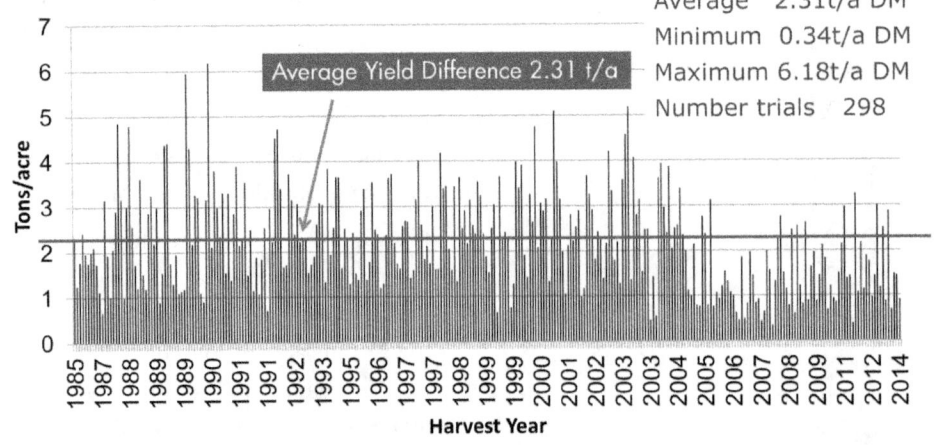

Average   2.31t/a DM
Minimum 0.34t/a DM
Maximum 6.18t/a DM
Number trials   298

Average Yield Difference 2.31 t/a

Advanced Techniques

Getting the most from reduced lignin varieties

Lignin occurs in plant cell walls to bind to the hemicellulose and fill spaces between cellulose and other molecules. It provides strength to plant cell walls to allow the plant to stand and seals the wall to allow the plant vascular system to transport water from the root to the leaves without leakage. Lignin also sequesters atmospheric carbon into vegetation.

On the other hand, lignin is not digestible in the animals and, since it binds to cellulose and hemicellulose, is a barrier to their digestion. Thus we want the minimum lignin necessary for plant function but, no more, so that the plant has the highest possible digestibility.

Recently alfalfa varieties have been released with reduced lignin content. Research has shown that a 10 to 15% lignin reduction results in 5 to 10% increased fiber digestibility with reduced undigestible NDF and greater rate of passage resulting in a gain of 2 lb milk/day. Such quality improvements would also be expected to show a 20 to 30% improvement in averge daily gain of growing animals. Alfalfa with smaller lignin reductions will have lesser changes in fiber digestibility and animal performance.

The lignin reduction can benefit farmers by:

• Producing higher quality forage, or

• Allowing later harvest for higher yield with the same forage quality as a standard variety, or

• Greater harvest flexibility.

Some benefits to farmers will be to have some fields with standard lines and some with the reduced lignin lines to spread the spring harvest window or to have all reduced lignin varieties to either produce higher quality forage or to produce increased yield from delayed cutting (and, in some cases) fewer cuttings.

When delaying harvest farmers should consider the additional needs for foliar fungicide and insect control.

• corn following alfalfa yields approximately 10% more than corn following corn, and

• corn rootworm is less of a problem in the first year following alfalfa than in corn following corn.

Winterhardiness

Winterhardiness is a measure of the alfalfa plant's ability to survive the winter without injury. It is measured on a scale of 1 to 6 with 1 being the most hardy and 6 being the least hardy. Winter-injured plants may survive the winter, but the buds formed in the fall for spring regrowth may be killed. Such plants have fewer shoots for first cutting and produce a lower yield.

Fall dormancy

Fall dormancy is measured by determining how tall alfalfa grows in the month following a September 1 cutting. More dormant types, such as Vernal, will remain short and low yielding through the fall period no matter how good the growing conditions are. Less-dormant varieties typically yield more in the fall, green up earlier in the spring, and recover more quickly between cuttings.

Plant breeders have finally broken the relationship between winterhardiness and fall dormancy. Until recently, obtaining higher yields meant selecting a variety with less dormancy and lower winterhardiness. The strategy now should be to choose less-dormant varieties that meet your winter survival requirement. These plants will green up earlier in the spring and recover more quickly between cuttings to give higher total season yields.

Disease resistance

Diseases may kill seedlings, reduce stand density, lower yields, and shorten stand life. The best disease management strategy is to select varieties with high levels of disease resistance. Determine potential for diseases on your farm and select alfalfa varieties with resistance to as many of them as possible. Knowing which diseases have occurred in your fields will help you choose varieties with the appropriate resistance. Look over the descriptions and pictures in the disease section of this guide, learn to recognize them, and select resistant varieties if the disease is occurring in your field. To estimate the potential for each disease to occur

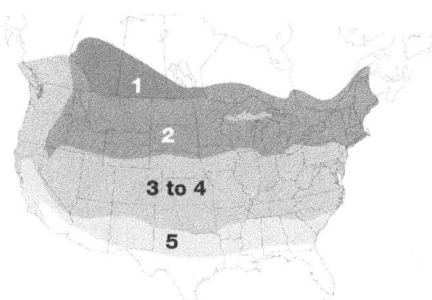

Winterhardiness needed. Varieties grown north of primary region of adaptation will suffer winterkill and injury. Western mountain valleys may grow less winterhardy varieties than indicated.

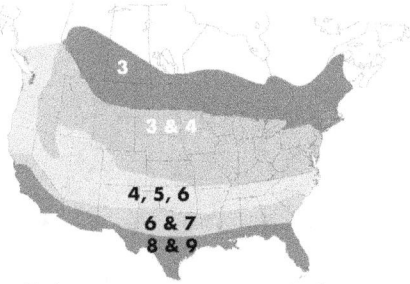

Fall dormancy recommended. Note that higher fall dormancy number indicates quicker recovery and should be used for silage or hay quickly removed from field.

in your area, refer to the maps in the disease management section.

Intended use

Most alfalfa is planted for harvest as hay or haylage with plans to keep stands as long as they are productive. Special situations may require different variety selection criteria. For example, when a short rotation is desired or when nitrogen for other crops is needed, yield is more important than persistence, so select varieties with high yields in the first 2 years. Also, varieties with higher fall dormancy (quicker recovery) may benefit in alfalfa harvested for silage but not for hay.

Planting

Time of seeding

Spring seeding is preferred over late-summer seeding in northern states due to a greater chance of successful stand establishment. Better growing conditions, such as a longer growing season, adequate soil moisture, and cool temperatures, enhance seed germination and establishment. Late-summer seeding is preferred in southern states because of the opportunity to establish alfalfa after growing another crop. Herbicides are not generally required for late-summer seeding. Irrigation allows late-summer seedings in all areas.

Spring seeding of alfalfa can begin as soon as the potential for damage from spring frosts has passed. At emergence, alfalfa is extremely cold tolerant. At the second trifoliate leaf stage (Figure 7), seedlings become more susceptible to cold injury and may be killed by 4 or more hours at 26°F or lower temperatures. Alfalfa seeded with a companion crop survives lower temperatures and longer exposure times before showing frost damage. Frost damage is usually not a problem by the time fields are tilled and ready to seed. Spring seedings have less weed competition and less moisture stress during germination than do late-summer seedings because of cooler temperatures.

The spring seeding dates shown on the map (Figure 8) are averages for the region. Seeding may be earlier on light soils, when a companion crop is used, or when forages are established using reduced-tillage or no-till methods. Irrigation may extend the seeding period later into the spring. Although successful stand establishment can be made outside the recommended dates, the likelihood of consistent success is low.

Successful late-summer seeding depends on soil moisture during the establishment period and sufficient plant growth before a killing frost (Figure 8). Do not seed unless good soil moisture is present. A preplant herbicide is usually not needed for light weed infestations because annual weeds will be killed by frost. Postemergence herbicides can be used if severe weed pressure or volunteer grain problems develop. Use of a companion crop is not recommended, especially if seeding before August 15, because it will compete with alfalfa for moisture. In many regions, Sclerotinia crown rot may be prevalent in late-summer seeding.

Alfalfa needs at least 6 weeks growth after germination to survive the winter. The plant will generally survive if it develops a crown before a killing frost. The crown allows the plant to store root reserves for winter survival and spring regrowth. Fields with less seedling development before a killing frost may have a greater problem with winter annual weeds, particularly in southern areas.

Figure 7. Alfalfa seedling development.

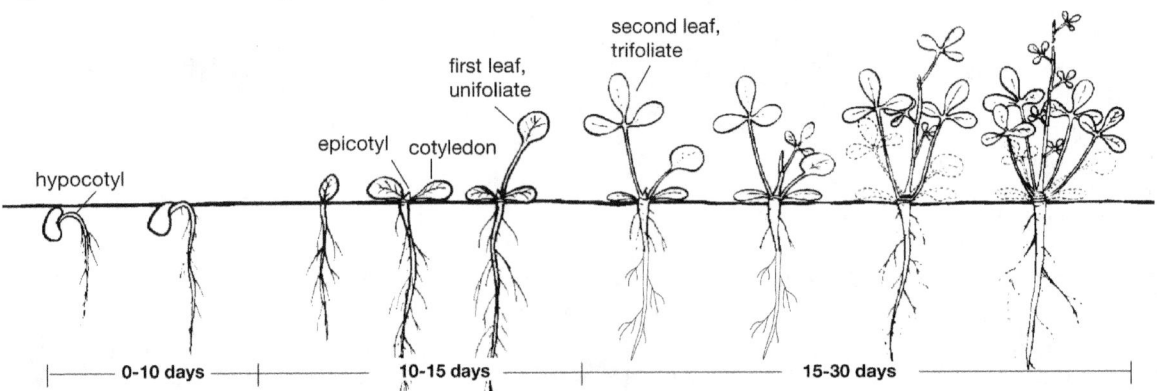

Source: Dodds and Meyer, North Dakota State University, 1984.

Figure 8. Spring and late-summer seeding dates.

■ **spring seeding preferred**

■ **late-summer seeding preferred**

spring seeding dates

May 1–30

April 15–May 15

April 1–30

March 15–April 15

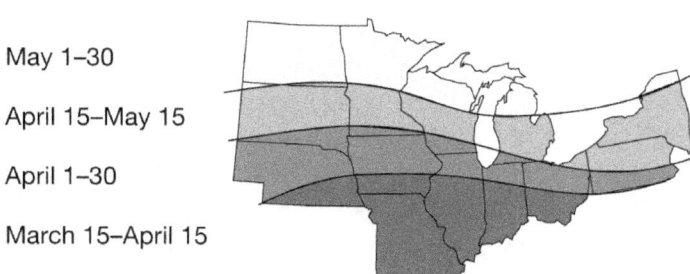

late-summer seeding dates

July 20–Aug. 1

Aug. 1–15

Aug. 15–Sept. 1

Sept. 1–15

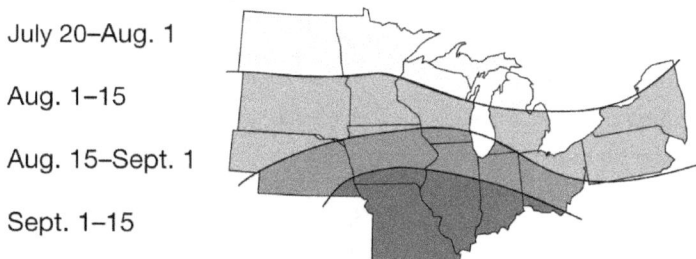

Minimizing competition from volunteer small grains or weeds is critical in northern regions to ensure adequate development of summer-seeded alfalfa prior to killing frost. Failure to do so cuts seedling establishment and lowers yields (Figure 9), particularly in no-till fields.

Field preparation

Field preparation should begin the year before seeding. Perennial weeds can be particularly competitive both during the seeding year and in subsequent years. Controlling weeds before seeding will help ensure a long-lasting, productive stand. Scout fields for perennial weeds and use appropriate control measures in the preceding crop. For example, if quackgrass is in a field where corn will be planted, use the appropriate combination of herbicide and tillage to control the quackgrass in the corn crop. Similarly, control Canada thistle, yellow nutsedge, dandelion, and other perennial weeds with an effective management program before seeding alfalfa. Be sure to follow herbicide replant restriction time intervals before seeding alfalfa to prevent herbicide carryover injury.

Figure 9. Effect of weed management systems for late-summer seedings on yield and stand the year following establishment.

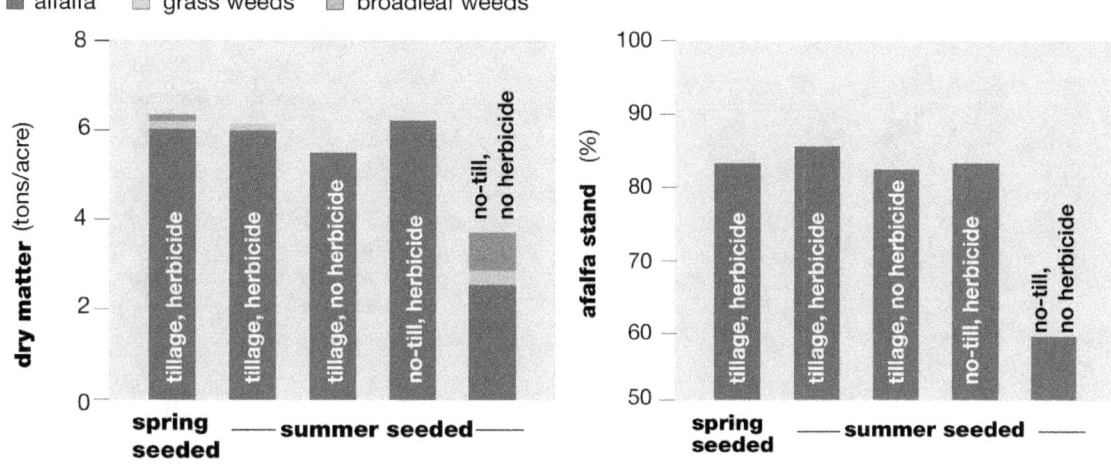

Source: Becker, University of Minnesota, 1993.

Conventional tillage practices vary from farm to farm but should consist of a primary tillage (moldboard plowing or chiseling) followed by disking. Primary tillage loosens the soil and helps control perennial weeds while disking controls weed regrowth, helps level the land, and breaks up large soil clods. The final tillage should be some type of smoothing operation. On level ground, primary tillage is best done in the fall as winter freeze-thaw cycles help break up clods. It also reduces field operations in the spring. On erosive soils, fall tillage may not be an option.

The ideal soil condition for conventional seeding should be a smooth, firm, clod-free soil (see picture) for optimum seed placement with drills or cultipacker seeders. Avoid overworking the soil as rainfall following seeding may crust the surface, preventing seedling emergence.

Seed inoculation

Rhizobium bacteria create nodules on alfalfa roots, allowing the bacteria to fix nitrogen where it becomes available to the plant. While many soils contain some Rhizobium bacteria from previous alfalfa crops, not all fields have

adequate numbers. To ensure the presence of the needed bacteria, purchase preinoculated seed or treat the seed using commercial inoculum available from seed dealers. Most varieties are preinoculated. These inoculant treatments often contain Apron fungicide as well, which protects against *Pythium* spp. and *Phytophthora* spp. diseases that reduce seedling emergence and kill young seedlings. Additionally areas with severe aphanomyces will benefit from planting seed additionally treated with the fungicide, Stamina. If treating seed yourself, make sure the inoculant was stored in a cool place before and after purchase; apply with a sticker— an adhesive compound to attach the Rhizobia to the seed—and thoroughly mix inoculum and seed before planting.

Seeding depth and rate

Alfalfa is a small-seeded crop and correct seeding depth is very important. Seed should be covered with enough soil to provide moist conditions for germination while allowing the small shoot to reach the surface (Figure 10). Optimum seeding depths vary depending on soil types. Plant

Soil should be firm enough at planting for a footprint to sink no deeper than 3/8 inch.

Advanced Techniques

Determining if seed is good quality

Look at the germination tag on the bag of seed to determine seed quality. The age of seed has little to do with establishment; all that counts is the ability to germinate.

Do the following:

1. Check the date of the germination test; it should have been run within the past year (preferably last 6 months).

2. Look at the % Germination; it should be above 90%. Note that % Germination includes the % hard seed (e.g., 90% germination with 15% hard seed means that 75% germinated and 15% were hard seed). Percent hard seed generally declines as the seed ages. Most hard seed of alfalfa will germinate within 30 days of seeding.

3. Look at the weed seed and crop Seed; the two should sum to less than 1%.

All seed should also be inoculated with *Rhizobium* bacteria for nitrogen fixation and Apron, and possibly Stamina, treated to reduce stand loss to seedling diseases.

Figure 10. Alfalfa emergence from various seeding depths.

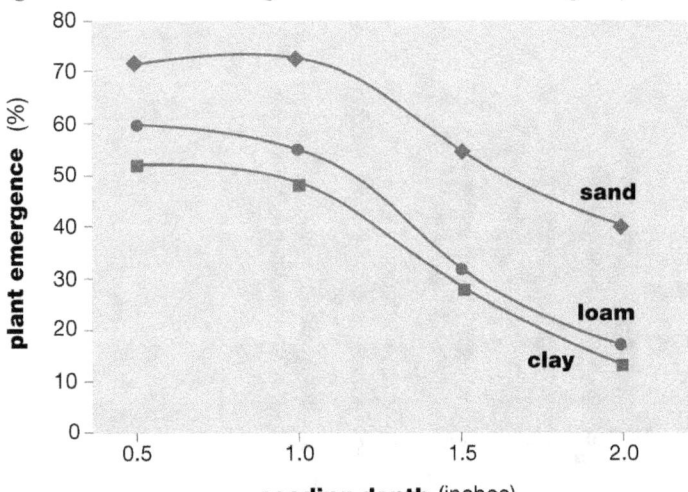

Source: Sund et al., University of Wisconsin, 1966.

Table 2. Effect of seeding rate on first-year alfalfa dry matter yields.

Seeding rate (lb/acre)	Dry matter yield (tons/acre)
12	3.4
15	3.6
18	3.6

Source: Buhler, Proost, and Mueller, University of Wisconsin, 1988.

seed 1/4- to 1/2-inch deep on medium and heavy textured soils and 1/2- to 1-inch deep on sandy soils. Shallower seedings may be used when moisture is adequate while deeper seedings should be used in drier soils.

Seeding rates should be between 12 and 15 lb/acre with good soil conditions and seeding equipment (Table 2). Higher seeding rates do not produce higher yields. Lower rates are normally used in arid regions. While these rates may be higher than needed for good stands under ideal conditions, the wide range of field conditions and environmental conditions experienced at seeding make this necessary to obtain consistently good stands. For example, extended periods of cool, wet weather can cause high seedling mortality (this can be reduced by planting Apron-treated seed). Hot, dry weather at seeding time likewise may reduce germination and seedling establishment. Under normal conditions, only about 60% of the seeds germinate and nearly 60 to 80% of the seedlings die the first year (Figure 11).

An important and often overlooked aspect of planting alfalfa is seeder calibration. Seed size can vary between varieties and between seed lots of the same variety. Calibrate seeding implements each time you use a new variety or a new seed lot, or if you use lime- or clay-coated seed. Regular calibration can help to avoid over- or underseeding.

Seeding with and without a companion crop

Direct seeding alfalfa (planting without a companion crop) allows growers to harvest up to two extra cuttings of alfalfa and produce higher quality forage in the seeding year as compared to alfalfa seeded with a companion crop. However, total forage tonnage may be less than that of companion-crop seedings. Some important management considerations are listed below:

- Select level fields with low erosion potential for direct seedings; use companion seedings where the erosion potential is greater. Erosive soils can be direct seeded using reduced-till or no-till methods that leave adequate residue on the surface.

- Effective weed management is critical in direct seeding (as no companion crop is present). See the section on weed management for details.

- Harvest the first cutting 60 days after germination, regardless of maturity stage. This eliminates many annual weeds and allows the second cutting to reach 10% bloom by September 1 in areas with short growing seasons.

Companion crops such as annual (Italian) ryegrass, oats, spring barley, and spring triticale help control erosion, reduce seedling damage from blowing sands, and minimize weed competition during establishment. Companion crops also provide additional forage when harvested as oatlage or grain.

Annual ryegrass provides a higher-quality first harvest and greater yield potential in the seeding year than small grain companion crops. Plant late-maturing annual ryegrass varieties, which remain vegetative in the seeding year (early-maturing types flower in the seeding year). Seed at 2 to 4 lb/acre.

For small grains, when companion crop growth is dense and grown to grain, alfalfa underneath is often damaged either by competition or by lodging of the small grain which smothers the alfalfa seedlings. Winter wheat, spring wheat, and rye usually compete too strongly with alfalfa seedlings and are less desirable as companion crops.

Following removal of spring-seeded companion crops, alfalfa regrowth and yield will be largely dependent on moisture availability. One or more harvests may be possible before the fall critical

Figure 11. Stand density during first 12 months (seeded at 12 lb/acre).

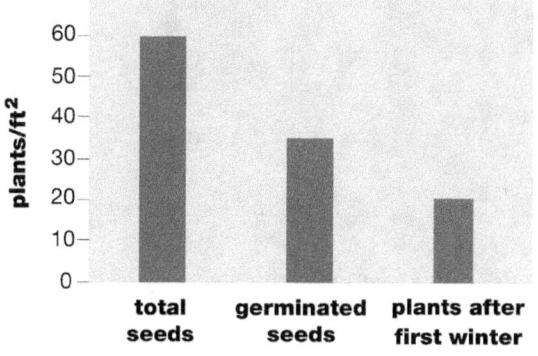

Source: Undersander, University of Wisconsin, 1995.

period. Usually, alfalfa regrowth competes well against summer annual weeds; however, herbicide use may be beneficial if weed populations are high and regrowth is slow.

In most of the alfalfa growing regions, companion crops are used only with spring seedings. With summer and fall seedings, moisture is often limiting and competition from the companion crop may limit alfalfa seedling development. If companion crops are needed with summer and fall seeding, the best strategy is to use the advanced technique described at the right.

For good alfalfa stands with companion seedings, manage the field to the advantage of the alfalfa rather than for the companion crop. Some important management considerations follow:

- Select companion crop varieties that are short, stiff strawed, and early maturing to avoid lodging and smothering the alfalfa.

- Seed companion oats or barley at 1 to 1 1/2 bushels/acre on heavy soils and 1 bushel/acre on sandy soils to reduce competition for light and moisture with the alfalfa seedlings.

- Limit nitrogen applications to no more than 30 lb/acre to avoid excessive competition and lodging of the companion crop.

- Harvest the companion crop at the boot stage rather than leaving it for grain. Harvesting at the boot stage reduces competition with alfalfa and minimizes the chance for lodging and smothering the alfalfa crop. This harvest stage provides optimum forage quality and yield of the companion crop.

- If you do harvest the companion crop for grain, cut it as early as possible to minimize lodging damage. Remove straw as quickly as possible to avoid smothering the alfalfa stand. Harvesting companion crops for grain is not recommended for good alfalfa stand establishment.

- If you plan to harvest the companion crop for grain, consider seeding an early variety in the spring and no-till seeding alfalfa into the grain stubble after harvest.

Seeding equipment

Many different types of drills and seeders are used to seed alfalfa. All will produce good stands when planting to an accurate seeding depth in a firm, moist soil. The keys to getting a good stand are placing the seed at 1/4 to 1/2 inches deep (slightly deeper on sands), covering the seed, and firming the soil around the seed.

Advanced Techniques

Getting direct seeding benefits while controlling erosion

Where the benefits of direct seeding are desired, yet the need for erosion control suggests a companion crop, it may be practical to seed oats as a companion crop and kill it with Poast Plus or Select herbicide, or Roundup if Roundup Ready Alfalfa was planted, when the oats are 4 to 6 inches tall. The oats will control weeds early, provide erosion control, and protect seedlings from wind damage. When the oats are 4 to 6 inches tall, spray with herbicide. After the oats have been killed, alfalfa will perform about the same as in a direct seeding (Figure 12). Thus, the erosion control benefits of a cover crop are achieved while still getting the higher alfalfa yield of a direct seeding. The practice may be particularly beneficial for fields with steep slopes or long gradients.

Figure 12. First-season yield and relative feed value (RFV) of alfalfa using different establishment methods.

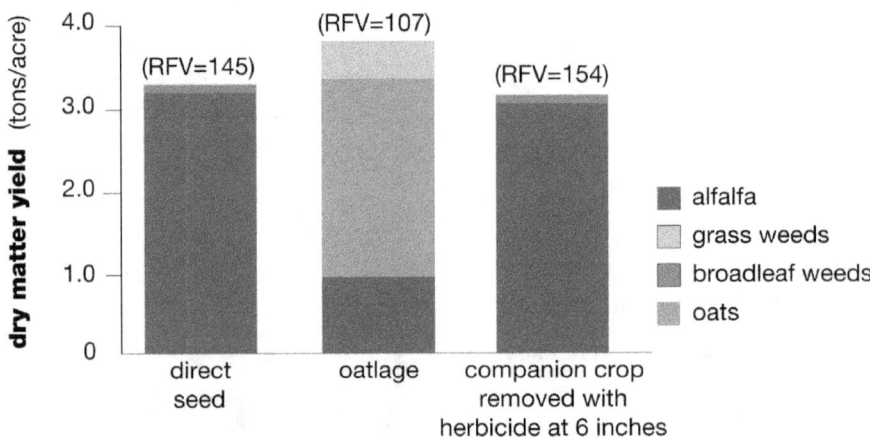

Source: Becker, University of Minnesota, 1989.

Cultipacker seeders, such as the Brillion seeder, broadcast the seed on the soil surface and then press it into the soil with rollers. These seeders have been a mainstay of alfalfa establishment because they give consistently even seed depth placement and good seed–soil contact for most soils. However, they do not work as well as drills on very hard ground or on very sandy soil.

Drills place the seed in rows, usually with 7- to 9-inch spacings, and can place fertilizer below the seed where it's most effective. To improve establishment, use press wheels mounted on the seeder or some other packing device, such as a cultipacker, pulled behind the seeder or used in a separate pass (Figure 13). The most common drills for forage establishment are grain drills that can seed a companion crop simultaneously with the alfalfa. Grain drills have poor depth control for seed placement. Drills adapted for forages have depth bands to overcome this problem. Alfalfa and companion crop seed must be put in separate seed boxes. Companion crops should be seeded 1 to 2 inches deeper than alfalfa. This can be done in a single pass by placing the drop tubes for the companion crop between coulters and for alfalfa behind coulters (see photo).

Figure 13. Importance of packing soil after seeding.

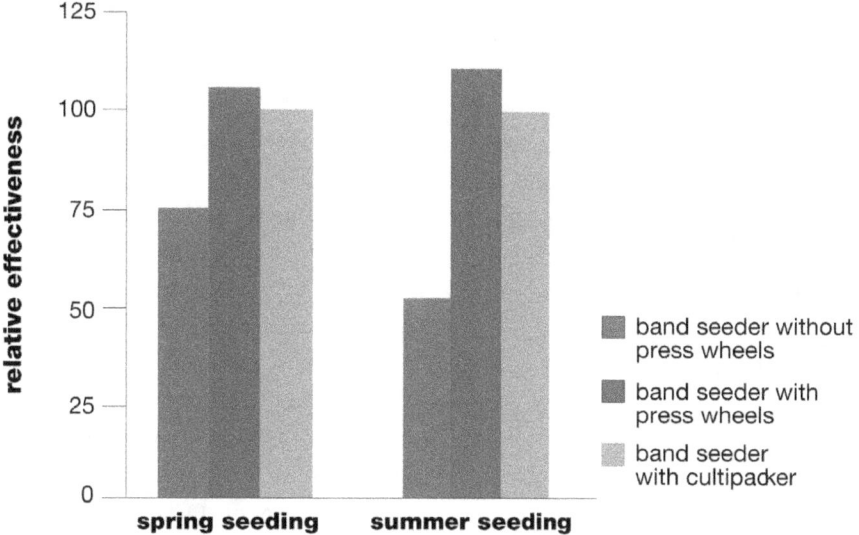

Source: Tesar, Michigan State University, 1984.

To plant companion crops 1 to 2 inches deeper than alfalfa in a single pass, place the drop tube for alfalfa behind coulters but before the packer wheel (a) and place the drop tube for the cover crop between coulters (b).

Reduced tillage and no-till planting

Due to the high potential for erosion on slopes using conventional tillage, a great deal of interest has been generated in reduced tillage alfalfa establishment. These practices include the use of a chisel plow rather than a moldboard plow, a single pass with a secondary tillage tool, or no tillage at all. Reduced tillage practices are generally successful when careful, timely management is used (Figure 14).

Crop residue management is an important factor in reduced-tillage seeding. Chisel plowing or disking typically chops the residue finely enough for conventional seeding implements to be used. In corn residue, a single disking may give the same result and level the field for easier hay making. Cultipacker seeders will not perform well with residue levels above 35% so a no-till seeder is recommended. Chopping stalks helps even the residue in the field and can reduce the amount of residue in the first alfalfa harvest.

Special attention must be given to weed management in reduced tillage systems. When direct seeding, weed control is more difficult as there is less tillage to decrease weed populations. Perennial weeds are the most difficult to control. Lack of deep tillage may give some perennial weeds a head start on the alfalfa. The use of a nonselective herbicide, such as glyphosate, to control perennial weeds (preferably in the previous fall) is critical prior to reduced or no-till seeding. Other weed control options are similar to conventional direct seeding and are discussed later in this publication. Oats can still be used as a companion crop.

Additional considerations in no-till alfalfa establishment are soil fertility and pH. As no tillage is done in the seeding year, materials that work best when incorporated, such as phosphorus fertilizers and lime, should be applied and worked into the soil before entering into no-till systems. If incorporation is not feasible, apply the finest grade of lime obtainable 1 to 2 years ahead of seeding to raise soil pH in the top inch of soil. (Lime moves

downward at about 1/2 inch per year on silt loam soils and somewhat faster on coarser soils.) Fine-grade alternative liming materials such as paper-mill lime sludge or cement plant kiln dust can also be used.

Many implement companies produce specialized no-till seeders for alfalfa and other crops. The design of these seeders differs among companies but should have the following features to ensure success:

- heavy down pressure,
- coulter ahead of disk openers to cut trash,
- double disc openers or an angled single disc opener,
- press wheels,
- small-seed box, and
- depth control mechanism.

Set seeding depths carefully as these implements are very heavy and may easily place seed deeper than optimum.

No-till seeders are often available for rent through Land Conservation offices, the USDA Natural Resources Conservation Service, or local fertilizer dealers and elevators.

Figure 14. Effect of seeding equipment on yield and stand in seedling year.

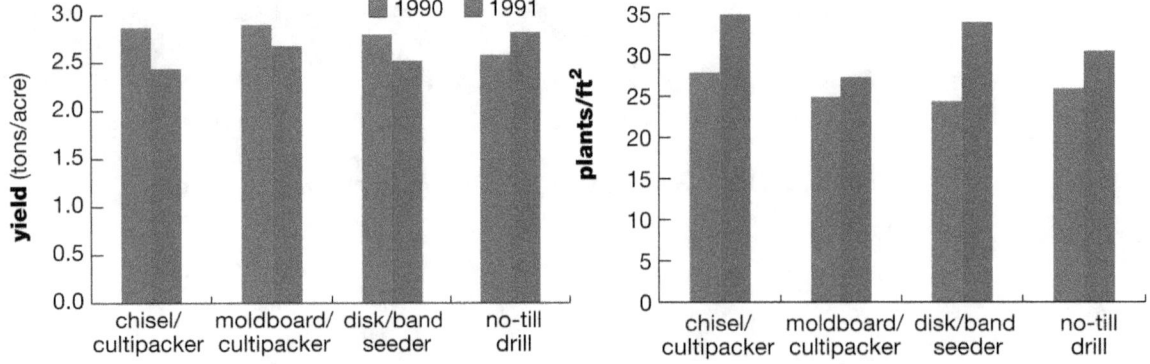

Source: Undersander and Mueller, University of Wisconsin, 1992.

Production

Once a good stand has been established, continued production and stand life depends on good management practices. Good management includes maintaining soil nutrients, applying manure judiciously, and controlling weeds and insects. Monitor diseases to estimate stand life and to determine resistance needed in future plantings. Finally, optimum production involves deciding when to rotate from stands that are no longer profitable.

Fertilize annually

Determine needs

Alfalfa has a relatively high demand for some nutrients compared to other commonly grown crops. Each ton of alfalfa dry matter harvested removes about 14 pounds of phosphate (P_2O_5) and 58 pounds of potash (K_2O). This is the nutrient equivalent of 150 pounds of a 0-10-40 fertilizer. Each ton of alfalfa also removes the calcium and magnesium found in about 100 pounds of aglime. See Table 3 for a complete list of nutrients removed. Since many of these nutrients are supplied from the native soil reserves, basing a fertility program on removals is not recommended.

Soil tests are the most reliable method for preventing nutrient deficiencies. Visual symptoms (Table 5 and photos) can be used to help assess nutrient needs for future yield. However, by the time visual symptoms appear on a crop, nutrient deficiency may be so severe that significant yield losses have already occurred. Visual symptoms can also reflect environmental conditions, restricted root growth, diseases or other problems not related to a soil nutrient shortage.

Plant tissue analysis can determine the nutritional status of your crop before any visual symptoms appear. While this method does not measure nutrient amounts for making a fertilizer recommendation, combining tissue analysis with a soil test makes for a comprehensive nutrient management system.

Advanced Techniques

Tissue testing

Tissue testing is very useful for assessing levels of sulfur and micronutrients. It can detect nutrient problems not easily detected with a standard soil test. Sample the top 6 inches of forage in bud to early flower stage in areas of the field that are free of other problems (insect, disease, drought, shade, etc.). Follow specific sampling and data interpretation guidelines to avoid misinterpretation. See Table 4 for a list of suggested sufficiency levels for the essential nutrients.

Some have recommended sampling from baled hay for tissue mineral analysis. While this can be useful, results are more variable than sampling as described in the first paragraph. Most importantly, be sure to use different sufficiency ranges than for samples from top six inches of forage.

Table 3. Pounds of nutrient removed per ton of alfalfa produced, dry matter basis.

Nutrient	Dry matter removed (lb/ton)
phosphorus (P)	6
phosphate (P_2O_5)	14
potassium (K)	48
potash (K_2O)	58
calcium (Ca)	30
magnesium (Mg)	6
sulfur (S)	6
boron (B)	0.08
manganese (Mn)	0.12
iron (Fe)	0.33
zinc (Zn)	0.05
copper (Cu)	0.01
molybdenum (Mo)	0.002

Table 4. Sufficiency levels of nutrients, top 6 inches of alfalfa at first flower.

Nutrient	Low	Sufficient	High
		— % —	
nitrogen	<2.50	2.50–4.00	>4.00
phosphorus	<0.25	0.25–0.45	>0.45
potassium	<2.25	2.25–3.40	>3.40
calcium	<0.70	0.70–2.50	>2.50
magnesium	<0.25	0.25–0.70	>0.70
sulfur	<0.25	0.25–0.50	>0.50
		— ppm —	
boron	<25	25–60	>60
manganese	<20	20–100	>100
iron	<30	30–250	>250
zinc	<20	20–60	>60
copper	<3	3–30	>30
molybdenum	<1	1–5	>5

Phosphorus deficiency

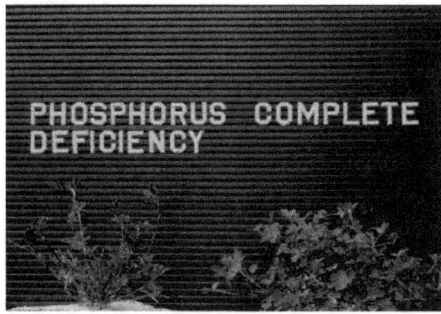

Deficient plants have blue-green leaves and stunted growth.

Leaflets often fold together, and the undersides may be red or purplish (left).

Potassium deficiency

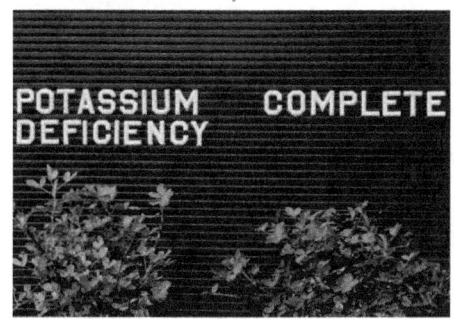

Leaves of severely deficient plants turn completely yellow.

Lower leaves of deficient plants are edged with white spots (left).

Sulfur deficiency

Stems are spindly with weak growth.

Leaves turn light green (left). Symptoms are similar to nitrogen deficiency.

Boron deficiency

Deficient plants have yellowed leaves on shortened stems.

Yellow coloring turns reddish to pruplish between veins.

Table 5. Nutrient deficiency symptoms for alfalfa.

Nutrient	Deficiency symptoms
nitrogen	Light green to yellow color, spindly growth.
phosphorus	Blue-green color, stiff, stunted and erect growth. Leaflets often fold together, and the undersides and stems may be red or purplish.
potassium	White spots around edge of leaf starting with lower leaves. In advanced cases leaves turn completely yellow and die.
calcium	Impaired root growth or rotting. Petioles collapse on youngest mature leaves.
magnesium	Interveinal chlorosis of lower leaves, margins initially remain green.
sulfur	Light green, similar to nitrogen deficiency, spindly stems and weak growth.
boron	Yellowing of leaves, shortened main stem growth between upper portion of shoots, dense top. Often confused with leafhopper damage.
manganese	Interveinal chlorosis of younger leaves.
iron	Interveinal chlorosis of youngest leaves, bleached appearance.
zinc	Reduced leaf size and upward curling of youngest leaves.
copper	Severe curvature of petioles, grayish spots in midleaf.
molybdenum	Pale green and stunted as with nitrogen deficiency.

Nitrogen

Alfalfa typically gets enough nitrogen from its symbiotic relationship with nitrogen-fixing Rhizobium bacteria and from soil organic matter, which releases nitrogen as it decomposes. On well-inoculated, established stands, topdressed nitrogen does not improve yields, quality, or stand vigor. Normally, adding nitrogen may lower yield and/or quality by stimulating growth of grasses and weeds. But in some cases, such as where soils have not been adequately limed, an application of 30 to 50 lb/acre of nitrogen can be used as a stop-gap measure to increase yields.

Phosphate and potash

Alfalfa needs relatively large amounts of phosphate and potash. Adequate phosphorus is important for successful establishment and good root development. Potash is essential for maintaining yields, reducing susceptibility to certain diseases, and increasing winterhardiness and stand survival. In the eastern portion of the Midwest, potassium is likely the most limiting nutrient to alfalfa production.

Phosphate and potash are relatively immobile when added to the soil. Phosphate bonds tightly on acidic clayey soils (pH < 5.5) and on very high pH soils (pH > 7.5) making it unavailable to plants. Potash can leach on some extremely sandy soils and on organic soils (peat or muck). Applications of phosphate and potash should be based on recommendations from a recent, well-calibrated soil test. Since alfalfa may take up more potassium than the plant needs, which creates animal health problems, do not topdress potash on soils testing very high for this nutrient. Forage tissue potassium levels should be monitored if luxury consumption of potassium is suspected.

Alfalfa absorbs most nutrients, including phosphate and potash, from the top 6 to 8 inches of soil. However, because phosphorus is immobile, alfalfa responds better to incorporated applications than to topdressed applications. Guidelines for annual phosphate and potash application include the following:

1. Apply topdress nutrients immediately after harvest and before regrowth resumes. Avoid contact with wet foliage.

2. Topdress following first cutting to stimulate second and third cutting regrowth or in early September to increase winterhardiness.

3. Avoid application when soils are soft (such as early spring) when physical damage to the alfalfa crown is likely.

4. Split the application to avoid salt damage if more than 500 lb/acre of material (irrespective of grade) is to be used in any year.

5. Base fertilizer purchases on cost per unit of plant food provided and need for all nutrients contained in fertilizer. For example, since there is no difference in nutrient availability with red versus white potash or with ortho- versus polyphosphate on most soils, the best choice is the least expensive product. Potassium-magnesium sulfate may be a superior potassium source where sulfur is needed and not supplied from another fertilizer material.

6. Foliar application should not be used for applying moderate to high rates of macronutrients, although it is an excellent method for applying micronutrients.

Advanced Techniques

Managing high potassium testing field

Excessive forage potassium levels in alfalfa (above 3%) can cause milk fever and other anion balance problems, especially for early-lactation cows. Where producers are concerned about forage potassium levels, the following management tips may help.

1. Use well-calibrated soil tests to guide potassium applications. The most severe forage potassium excesses were observed when high rates of potash were topdressed on soils that already tested excessively high for potassium.

2. Allow the alfalfa to mature a few days longer. As alfalfa matures, tissue calcium levels decrease.

3. Cut alfalfa as low as possible without damaging the crown and preserve leaves during harvest. Potassium is concentrated in the upper stems; therefore, including a higher percentage of lower stems and leaf tissue will lower forage potassium levels by dilution.

4. All forages contain about the same amount of potassium when grown under similar environmental conditions, so adding grasses or other forages will not lower potassium content.

Secondary nutrients

Calcium and magnesium deficiencies are very rare, especially where soil pH has been maintained in the desired range for alfalfa. Symptoms of magnesium deficiency appear when the soil test drops below 50 to 100 ppm magnesium. Magnesium can drop below that level on acidic sandy soils where repeated high amounts of potassium have been applied, on soils where only calcitic liming materials have been used, and on calcareous organic soils. The most economical way to avoid calcium or magnesium problems is to follow a good liming program with dolomitic limestone. Where soil pH is adequate and extra magnesium is needed, apply magnesium sulfate (epsom salts) or potassium-magnesium sulfate (Sul-Po-Mag or K-Mag) at 20 to 50 lb/acre magnesium per year.

Sulfur deficiencies are likely when high sulfur-demanding crops such as alfalfa are grown. Sulfur from precipitation has been reduced due to less industrial air pollution. Where 10 to 25 lbs/acre each year were generally received in rain in the past, most regions now receive less than 5 lbs/acre each year. Since alfalfa uses 20 lbs/acre or more sulfur per year, many regions need fertilization with sulfur that did not previously. The amount of sulfur in manure depends on the kind of animal manure (Table 6). Some subsoils, especially those that are acidic and clayey, may contain enough sulfur for high-yielding crops even though the plow layer may test low.

Where the sulfur need has been established, either elemental sulfur or sulfate forms can be used on alfalfa. Sulfate-sulfur is immediately available to the crop, whereas elemental sulfur must be biochemically converted to sulfate before it can be used, which is a slow process taking several months. When applied at 25 to 50 lb/acre, sulfate-sulfur will generally be adequate for 1 or 2 years of alfalfa production. In contrast, elemental sulfur applied at the same rate should last for the term of the stand. Elemental sulfur converts to sulfate more rapidly when incorporated.

Micronutrients

Plants need only very small amounts of micronutrients for maximum growth. While a deficiency of any essential element will reduce plant growth, overapplication of micronutrients can produce a harmful level of these nutrients in the soil that is difficult to correct, especially on coarse-textured soils. Soil tests are available for some micronutrients, but plant analysis is generally more reliable for identifying micronutrient problems.

Boron is usually the only micronutrient that is needed in a fertilizer program for alfalfa. Boron management depends on the texture of the soil. Sandy soils do not hold boron as tightly as clayey soils. A high test in a sandy soil may be only medium in a silt loam. For alfalfa where the soil test is very low or low on medium-textured soils, apply 2 to 3 lb/acre boron once in the rotation. On sandy soils apply 0.5 to 1 lb/acre boron each year. Due to the low rate of material needed, boron is often mixed with other fertilizers such as potash. Do not apply boron near germinating seeds.

Alfalfa has a relatively high requirement for molybdenum. However, since molybdenum availability increases as pH increases, liming to optimal pH levels usually eliminates molybdenum problems. Manganese, zinc, iron, and copper are rarely deficient in alfalfa. In special situations where deficiencies are suspected, contact your county Extension office or consultant before treating.

Table 6. Estimate of available sulfur from manure as affected by animal and manure type.

Animal	Sulfur content			
	Solid		Liquid	
	Total	Available	Total	Available
	———— lb/ton ————		——— lb/1000 gal ———	
dairy	1.5	0.8	4.2	2.3
beef	1.7	0.9	4.8	2.6
swine	2.7	1.5	7.6	4.2
poultry	3.2	1.8	9.0	5.0

Irrigation

Improper irrigation limits alfalfa yield more often than any other management factor in semi-arid areas. Water use is generally estimated as evapotranspiration (ET), the combined evaporated water from soil and plant surfaces. Alfalfa ET normally varies from 0.1 to 0.35 inches per day (Figure 1) producing a seasonal ET of 36 inches per year in the semi-arid Northwest and up to 72 inches seasonal water use in the Southwest due to a longer and hotter growing season.

Alfalfa yield is linear to ET but the actual relationship varies depending on location and especially with relative humidity. As relative humidity increases the ratio of inches of ET per ton decreases. Water use efficiency is highest when the water supplied to plants approximates ET. For much of the semi-arid west it takes about 6.6 inches of water to produce each ton of alfalfa. This assumes 85 percent irrigation efficiency or 15% of water evaporated or run off before infiltrating the soil.

Water for growth can come from stored soil water, irrigation and rain. In heavier soils, water in the soil profile from the previous fall's irrigation and winter and spring precipitation will reduce irrigation water needed during the season. Stored soil water may be crucial to high yields because of low water infiltration rates in heavy soils. Sandy soils have much less water-holding capacity but have higher water infiltration rates.

Plant stress often occurs when available soil moisture falls below 50 percent and plants are unable to transpire adequate water to grow at an optimum rate. Lost yield can never be "made up" by irrigating more than necessary following the stress! Saline soils can also cause plant stress and may be a result of not enough water applied to leach salts below the root zone, or from irrigation water with high salt content. Alfalfa breeding programs are making progress developing varieties with more seedling salt tolerance and forage production in saline soils.

Border, corrugation (furrow), controlled flooding, and sprinkler irrigation can be used on alfalfa. Choose the method best suited to your slope, soil, water supply, and labor supply.

Irrigation scheduling is best accomplished by the water balance method, in which water inputs equal outputs, can be used to estimate the soil moisture condition. Most weather stations report a reference ET which is adjusted with a crop coefficient for each individual crop. Use estimated water consumption provided by services such as AgriMet for irrigation scheduling where possible: http://www.usbr.gov/. Use soil moisture sensors and a soil probe or shovel to check your soil moisture and verify the actual field conditions. The root zone should be filled with moisture just before the period of peak crop water use.

Irrigation scheduling principles:

1. Begin season with full soil water profile.

2. Monitor the soil profile for moisture content. Irrigate early to fill the root zone. This is important to have healthy roots in deep soil to take advantage of the soil's water holding capacity.

3. The soil surface should be dried to minimize soil compaction during harvest. The interval between irrigation and harvest varies from two days in lighter textured sandy soil at high summer ET rate, to 13 days in heavier clay soils at low ET rates. The soil water reserve can be used for alfalfa growth when irrigation is halted for harvest or when application rate does not keep up with ET.

4. Begin irrigating again as soon as harvest is removed to refill soil profile. Stress during early regrowth will severely limit the next crop yield.

Figure 15. Average evapotranspiration (ET) by alfalfa cut four times. Alfalfa was irrigated at maximum pivot capacity (6.5 gallons per minute, 85% efficiency). Note that in mid-summer the amount applied is less than the amount lost by ET. This means the alfalfa must use soil moisture reserves or suffer reduced yield.

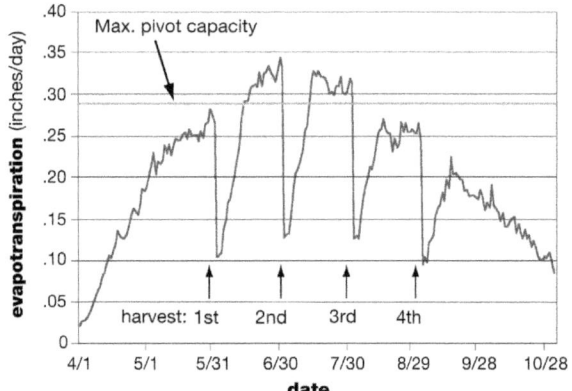

Manure management

Manure is a complete nutrient source, containing all of the major nutrients, secondary nutrients, and micronutrients. In addition, manure promotes biological activity in the soil and enhances the soil physical properties. While manure may be beneficial to soil, applying manure on alfalfa fields can create problems. Manure can burn leaves, reducing yield and quality. The mechanics of applying manure can compact soil and damage crowns, which in turn lowers yields and shortens stand life. Also, nitrogen in manure can stimulate weed and grass growth.

If possible, spread manure on other crops that can benefit from the nitrogen. Alfalfa will use applied nitrogen but does not need it due to its ability to fix nitrogen. When too much manure and/or too little cropland force application of manure to alfalfa, top management practices are required.

Use the following guidelines to minimize alfalfa damage when applying manure to the alfalfa stand:

1. Choose fields that have the most grass, usually the oldest stands, since these will benefit most from nitrogen in manure.

2. Apply no more than 3,000 gallons of liquid manure or 10 tons of solid manure per acre. Applying more may cause salt burn, and damage or suffocate plants. Use supplemental fertilizer if additional nutrients are required.

3. Spread manure immediately after removing a cutting so manure contacts the soil instead of the foliage. This reduces the risk of salt burn and minimizes palatability problems.

4. Adjust the spreader to break up large chunks of manure that can smother regrowth.

5. Spread manure only when soils are firm to limit soil compaction and to avoid damaging crowns.

Weed management

Weeds reduce alfalfa production during establishment by competing with and choking out young alfalfa seedlings. Weeds also invade established alfalfa fields and reduce forage quality and alfalfa yield. Effective weed control begins before seeding and continues throughout the life of the stand. The most important factor in weed management is to establish and maintain a vigorous alfalfa crop. Proper soil fertility and pH, seedbed preparation, varietal selection, and appropriate cutting schedules cannot be overemphasized to prevent weed encroachment. If using a herbicide, remember that application timing and rates vary. *Always read the product label for application instructions.*

Weed management before planting

Most alfalfa stands are left in production for several years. The absence of tillage during the life of the stand naturally favors invasion by perennial weeds. It is very important to eliminate perennials before establishing alfalfa. Herbicides for perennial weed control may be applied in spring or fall. Fall application is recommended in most cases for more consistent control.

Waiting to apply nonselective herbicides to perennial weeds at the proper growth stage in spring may delay alfalfa planting past the optimum time unless Roundup Ready alfalfa is planted.

Herbicides for perennial weed control the year before seeding alfalfa include dicamba, glyphosate, 2,4-D, Stinger, Permit, and tank mixes of these herbicides. Carryover from dicamba and Stinger will damage alfalfa seedlings unless they are used far enough in advance of alfalfa planting. Consult labels for specific plant back recommendations.

One of the most serious perennial weed problems in alfalfa stands in northern states is quackgrass. Fall application of glyphosate is more effective than spring application. Quackgrass should be actively growing when glyphosate is applied.

Weed management in the seeding year

Tillage is an important part of a weed management program when establishing alfalfa. Thorough tillage helps uproot existing annual weeds and sets back established perennial weeds. Final tillage should be done as near planting as possible to allow alfalfa a head start on weed growth.

Herbicides can improve establishment, especially in fields with high weed densities. Volunteer plants from the previous crop must be controlled in fall seeded alfalfa. As little as one wheat plant per square foot will reduce the alfalfa stand. Several herbicides are currently labeled for use in new alfalfa seedings. This section describes the most commonly used options. Performance ratings for each herbicide are listed in Table 7.

Direct-seeded plantings: Preplant-incorporated treatments

Eptam (EPTC) is a preplant-incorporated herbicide that controls annual grasses and several broadleaf weeds. Eptam must be thoroughly incorporated to a depth of 2 to 3 inches. Incomplete incorporation may cause streaking and alfalfa injury or loss of herbicide. Incorporate, working the field in two different directions. Eptam can temporarily stunt alfalfa and the first leaves may not unfold properly. Injury may be more pronounced when applied under cool, wet weather, when high rates have been applied, or when poorly incorporated. Do not use Eptam if any atrazine was used in the previous 12 months as severe injury may result. Do not use Eptam if planting a forage grass crop with alfalfa as grass seedlings will be killed by Eptam.

Treflan (trifluralin) is a preplant-incorporated herbicide that controls annual grasses and some annual broadleaf weeds. Treflan will not control grass plants growing from rhizomes, such as quackgrass. Treflan must be incorporated to a depth of 2 to 3 inches. Incorporation may be delayed for up to 24 hours, but prompt incorporation is best. Incorporate with a tillage implement according to label directions. Do not use Treflan if planting a forage grass crop with alfalfa as grass seedlings will be killed by Treflan. Injury rarely occurs from Treflan applied at recommended rates.

Direct-seeded plantings: Postemergence treatments

Buctril (bromoxynil) is a postemergence contact herbicide that controls many common broadleaf weeds. For best results, treat when alfalfa has at least four trifoliate leaves and when weeds are 2 inches or less in height and have no more than four leaves. Buctril gives fair to good pigweed control if plants are small and actively growing when applied. Serious alfalfa injury may occur if the temperature exceeds 70°F within 3 days after application. Do not treat alfalfa stressed by moisture, insect injury, or other causes. Treated

Table 7. Alfalfa tolerance and herbicide effectiveness in direct seedings. *(Buctril is the only herbicide registered for use on alfalfa seeded with companion crops.)*

	Preplant incorporate		Post emergence							
	Eptam (EPTC)	Treflan (tifluralin)	Glyphosate	Buctril (bromoxynil)	Butyrac (2,4-DB)	Poast Plus (sethoxydim)	Prowl H2O (pendimethalin)	Pursuit (imazethapyr)	Raptor (imazamox)	Select (clethodim)
Alfalfa tolerance	F/G	G	E	F/G	G	E	E	G	G	E
Grasses										
barnyardgrass	G/E	G/E	E	P	N	G/E	G/E	G	G	G/E
foxtails	G/E	G/E	E	P	N	E	G/E	G	G/E	E
quackgrass	P/F	P	G/E	N	N	F/G	P	P	P/F	G
wild oats	G	F	F	N	N	G/E	P	F	F	E
Broadleaves										
eastern black nightshade	F	P/F	E	G/E	F	N	P	E	E	N
hoary alyssum†	F/G	P	G	F	F	N	P	F	F	N
kochia	F	G	G/E	F/G	F	N	G	G	G	N
lambsquarters	F/G	F/G	G/E	G/E	G/E	N	G/E	F/G	G	N
night-flowering catchfly	P	P	E	F/G	P	N	P	P	P	N
pigweed spp.	F/G	G/E	E	F	G/E	N	G/E	G/E	G/E	N
ragweed, common	F/G	P/F	P	G/E	G/E	N	P	F	F/G	N
smartweed spp.	P	P/F	G/E	G	P	N	P	G	G	N
velvetleaf	F/G	N	G/E	G	G/E	N	N	G/E	G/E	N
wild mustard	P/F	P/F	E	G	F	N	N	G/E	G/E	N

Abbreviations: E = excellent; G = good; F = fair; P = poor; N = no control; -- = no dataacontrol ratings for annual seedlings only.
Source: Adapted from Renz, University of Wisconsin, 2010.
† Control ratings for annual seedlings only.

fields cannot be harvested or fed for 30 days after application.

Butyrac (2,4-DB) is a postemergence systemic herbicide that controls many annual broadleaf weeds but is weak on larger mustards and smartweed and will not control grasses. It suppresses some perennial broadleaf weeds. Apply when seedling weeds are small and actively growing. Correct timing is critical as control is less effective on larger weeds. Check the label for specific rates according to weed species and size. Treated forage cannot be harvested or grazed for 60 days after application.

Buctril can be tank-mixed with Butyrac to improve control when weeds in the mustard or smartweed family (which are sensitive to Buctril), and pigweed (which is more sensitive to Butyrac) are present. Forage treated with this combination cannot be harvested or fed for 60 days after application.

Glyphosate kills a wide range of grass and broadleaf weeds, and Roundup Ready Alfalfa has excellent tolerance to glyphosate. In Roundup Ready Alfalfa, apply glyphosate when weeds are 4 to 5 inches tall as this will provide effective weed control. It is also recommended to treat alfalfa when it has 3 to 4 trifoliate leaves as this will eliminate the small percentage of alfalfa seedlings that do not contain the resistance gene (<10% of seed). Typically these two events occur at the same time in the field, so only one application is required. Fields cannot be harvested for 5 days following application.

Poast Plus (sethoxydim) and Select 2 EC (clethodim) are selective postemergence systemic herbicides that control most annual grasses present and suppress perennial grasses (including quackgrass) in alfalfa. Apply to annual grasses at the heights indicated on the labels. Grasses must be actively growing for best results.

Poast Plus and Select can be tank-mixed with Butyrac and applied to newly seeded alfalfa to control a mixture of grass and broadleaf weeds. With a tank mix, the possibility of crop injury increases because the oil concentrate increases Butyrac uptake. Use the rate of product as indicated for the weed species present. Do not add liquid fertilizer solution or ammonium sulfate when tank-mixing with Butyrac. Treated forage cannot be harvested or grazed for 60 days following application. It may be difficult to apply this tank mix at the proper time to adequately control both grasses and broadleaf weeds because each may not be at the best stage for control at the same time.

Prowl H2O (pendimethalin) is a preemergence herbicide that is effective at controlling many small seeded grasses and broadleaf weeds and would be a good fit for thinning stands with many annual weeds. Apply to established fields before weed emergence when alfalfa is less than 6 inches tall. Its effectiveness for winter annuals is variable unless higher rates are utilized. Fields cannot be harvested for 28 to 50 days following application to established alfalfa depending on the rate applied. Consult the label for further information.

Pursuit (imazethapyr) can be applied postemergence when seedling alfalfa has two or more trifoliate leaves and the majority of the weeds are 1 to 3 inches in height. Pursuit controls many

Advanced Techniques

Roundup Ready Alfalfa

Roundup Ready alfalfa is a powerful tool in the growers' arsenal. Weeds can now be effectively controlled in new seedings without the constraints of other herbicides, such as needing to allow alfalfa to grow to the 5-leaf stage, narrow windows of application, relatively long pre-harvest intervals, risk of crop injury, requirement for soil incorporation, and/or narrow weed control spectrum.

Roundup Ready alfalfa makes it easier for farmers who want pure alfalfa stands but need to establish with a cover crop. They can seed the alfalfa with oats (1 bu/a) or italian ryegrass (2 to 4 lb/a) and then apply glyphosate when the oats or ryegrass is 6 to 8 inches tall. This practice provides the benefits of reduced wind and water erosion and early weed control until the alfalfa is established and maintains the yield potential of the direct seeding method.

Alternatively, the cover crop can be grown out, harvested, and then the field sprayed with Roundup.

Roundup Ready alfalfa allows more flexibility and cost effectiveness when controlling weeds in established stands. Glyphosate controls a broader spectrum of weeds than most other herbicide programs, especially controlling winter annual and perennial weeds.

When rotating Roundup Ready alfalfa fields to other crops, use tillage and/or a herbicide such as 2,4-D, dicamba, or clopyralid in the fall after the final alfalfa harvest.

annual grass and broadleaf weeds and suppresses some perennial weeds. Pursuit can be tank-mixed with Buctril, Butyrac, or Poast Plus. Use a labelled adjuvant and a liquid fertilizer solution such as 28% nitrogen or 10-34-0 or ammonium sulfate to the spray solution. Following application, you must wait 30 days before grazing or harvesting and 4 months before replanting alfalfa back into the stand. Allow 60 days before grazing or harvesting if Butyrac is included.

Raptor (imazamox) is similar to Pursuit in both its chemistry and use guidelines. When applied to alfalfa with two or more trifoliate leaves and to weeds that are less than 3 inches tall, Raptor controls the same weeds as Pursuit with improved control of common lambsquarters and foxtail species. There is no waiting period for harvest of alfalfa after Raptor application.

Companion-crop seeded plantings

Buctril (bromoxynil) can be used in companion seedings to control several broadleaf weeds. It is effective on wild mustard and common lambsquarters. Buctril may cause serious alfalfa injury if the temperature exceeds 70°F within 3 days after application and if the alfalfa has fewer than four trifoliate leaves. Fields may be harvested 30 days after application.

Weed management in established alfalfa

Weeds encroach on alfalfa as stand growth slows due to poor fertility, disease and insect problems, and winter injury. Removing weeds from alfalfa seldom increases the tonnage of harvested forage. Rather, the proportion of alfalfa in the harvested forage increases. Whether this affects forage quality depends upon the weed species and their stage of growth. Dandelions and white cockle, for instance, do not influence forage quality and animal intake while weeds such as yellow rocket and hoary alyssum are unpalatable and decrease animal intake. The higher fiber content of grassy weeds also decreases intake. Refer to Table 8 for a comparison of the relative impact of weeds on forage quality.

The decision to use herbicides for weed control in established alfalfa stands should be based on the degree of the weed infestation, the type of weeds present, and most importantly, the density of the existing alfalfa stand. Alfalfa stands 3 years or older should have at least 55 stems per square foot. For treatment to be economical, weed infestations must be severe enough and of species that reduce forage quality, and alfalfa stand density must be high enough to respond to the decreased competition upon weed removal. Alfalfa does not spread into open areas, so removing weeds in thin stands often means weed reinfestation. The cost of herbicide treatments such as Velpar and metribuzin can generally be spread over 2 years because weeds will be suppressed for that length of time.

Table 9 compares the herbicides available for established stands. The following information describes the herbicides and when to apply them.

Table 8. Impact of common weeds on forage quality.

| | Relative seriousness | | |
	Serious	Moderate	Slight
annual weeds	cocklebur	green foxtail	lambsquarters
	Eastern black nightshade	pennycress	pigweeds
	giant foxtail	shepherd's purse	ragweed, common
	giant ragweed	velvetleaf	
	smartweeds		
	yellow foxtail		
perennial weeds	curly dock	Canada thistle	dandelion
	hoary alyssum	quackgrass and other grasses	white cockle
	yellow rocket		

Source: Doll, University of Wisconsin, 1998.

Butyrac (2,4-DB) may be applied to established stands to control several broadleaf weeds but is weak on larger mustards and smartweed and will not control grasses. It gives some suppression of perennial broadleaf weeds. Apply when seedling weeds are small and actively growing. Correct timing is critical as control is less effective on larger weeds. Check the label for specific rates according to weed species and size. Treated forage cannot be harvested or grazed for 30 days after application.

Chateau can be applied to established alfalfa (previously harvested) any time alfalfa is less than 6 inches tall. Make applications before emergence of target weed species as Chateau has only preemergence activity on annuals and germinating perennial weeds that are historically difficult to control once established. Fields cannot be harvested

or grazed within 25 days of applications. Chateau can be impregnated onto dry fertilizer for simultaneous application. Do not add any adjuvant or product formulated as an emulsified concentrate (EC).

Metribuzin (formerly sold as Sencor) controls a broad range of annual and perennial weeds, including fair to good control of dandelion and quackgrass. Alfalfa must be established for at least 12 months before using Sencor. To avoid injury, apply in early spring after the ground is thawed and while alfalfa is still dormant, or impregnate the herbicide onto dry fertilizer and apply when alfalfa is less than 3 inches tall and the foliage is dry. Rates vary with soil type and weed infestation. Consult label for appropriate rates as well as for crop rotation restrictions. Treated alfalfa may be harvested or grazed 28 days after application.

Poast Plus (sethoxydim) or Select (clethodim) may be applied to established stands of alfalfa to suppress quackgrass or control annual grasses. Treat when quackgrass is 6 to 8 inches tall and when annual grasses are small and actively growing. Do not apply to grass-legume mixtures as forage grasses will be stunted or killed. Alfalfa treated with Poast Plus may be harvested after 7 days as green chop or haylage and after 14 days for dry hay. If treated with Select, alfalfa can be harvested, fed, or grazed after 15 days. Treatment can be applied in spring or after any harvest during the summer.

Prowl H2O (pendimethalin) is a preemergence herbicide that is effective at controlling many small seeded grasses and broadleaf weeds, and would be a good fit for thinning stands with many annual weeds. Apply to established fields before weed

Table 9. Alfalfa tolerance and herbicide effectiveness on common weeds in established stands.

	Chateau	Butyrac (2,4-DB)	Glyphosate	Poast Plus (sethoxydim)	Prowl H2O (pendimethalin)	Select (clethodim)	Pursuit	Metribuzin	Raptor	Velpar (hexazidone)
Alfalfa Tolerance	G/E	F/G	E	G	E	G	G	G/G	G	G/G
Annual Weeds										
field pennycress	E	F/G	E	N	P	N	F	G	E	G
foxtail spp.	G/E	N	E	G	G/E	G	F/G	G	G/E	G
night-flowering catchfly	–	P	–	N	–	N	–	G	–	G
shepherd's purse	E	F/FG	E	N	F/G	N	G/E	G	G/E	G
Virginia pepperweed	–	F/G	–	N	–	N	–	G	–	G
Biennial Weeds										
spotted knapweed	P	F	G/E	N	P	N	–	F	–	N
Perennial Weeds										
Canada thistle	P	P	G/E	N	P	N	P	P	P/F	N
curly dock	P	P	–	N	P	N	P	F	P/F	F
dandelion	P	P	G	N	P	N	P	F/G	P/F	F/G
hemp dogbane	P	N	E	N	P	N	P	P	–	N
hoary alyssum	P	F	–	N	P	N	P	F/G	F	G
orange hawkweed	P	N	–	N	P	N	–	P	–	N
quackgrass	P	N	G/E	F/G	P	F/G	P	F/G	P/F	F/G
sowthistle, perennial	P	P	E	N	P	N	P	N	G	N
white cockle	P	P	F/G	N	P	N	P	F	P/F	F
wirestem muhly	P	N	E	F/G	P	F/G	P	P	P/F	F
yellow rocket	P	P	G/E	N	P	N	F/G	F/G	F/G	G

Abbreviations: E = excellent; G = good; F = fair; P = poor; N = no control; -- = no data.
Source: Adapted from Renz, University of Wisconsin, 2010.

emergence when alfalfa is less than 6 inches tall. Its effectiveness of winter annuals is variable unless higher rates are utilized. Fields cannot be harvested for 28 to 50 days following application to established alfalfa depending on the rate applied. Consult the label for further information.

Pursuit (imazethapyr) can be applied to established alfalfa for pre and postemergence control of annual weeds. Apply in the spring or fall to dormant alfalfa or after a cutting before regrowth exceeds three inches. Often adjuvants can improve performance of this product; consult the label for additional information. Good coverage is essential for adequate weed control, and weeds treated after a recent harvest may not receive much herbicide and be inadequately controlled. Fields cannot be harvested or grazed within 30 days of applications.

Raptor (imazamox) can be applied to established alfalfa for pre and postemergence control of annual weeds similar to Pursuit. Apply in the spring or fall to dormant alfalfa or after a cutting before regrowth exceeds three inches. Often adjuvants can improve performance of this product; consult the label for additional information. Good coverage is essential for adequate weed control, and weeds treated after a recent harvest may not receive much herbicide and be inadequately controlled. Fields cannot be harvested anytime after application.

Velpar (hexazinone) controls a broad spectrum of annual and perennial weeds, including fair to good control of dandelion and quackgrass. Alfalfa should be established for 1 year or more prior to treatment. Apply Velpar in spring to dormant alfalfa or before new growth exceeds 1 to 2 inches. Treating taller alfalfa will severely injure plants. Rates vary according to weed types present and soil type. Consult the label for specific recommendations. Do not graze or feed treated hay for 30 days. Corn may be planted 1 year after treatment; for all other crops, including alfalfa, you must wait 2 years before planting.

Fall-applied Velpar controls certain species (especially winter annuals) but is less effective on dandelion than spring applications. The uncertainty of winter survival of alfalfa also makes fall treatment a risky venture in most areas.

Disease management

Several diseases occur in alfalfa stands that can kill seedlings, limit yields, and shorten stand life. The occurrence and severity of diseases depends on environmental conditions, soil type, and crop management. Few economical control options are available for diseases once they're present in a field, but knowing which diseases are present can help you select resistant varieties for future plantings.

Anthracnose

Anthracnose occurs most often under warm, moist conditions and causes yield losses of up to 25%. On susceptible plants, stems have large, sunken, oval- to diamond-shaped lesions. Large lesions are straw colored with brown borders. Lesions can enlarge and join together to girdle and kill one to several stems on a plant. Girdled stems may wilt suddenly and exhibit a "shepherd's hook." This should not be confused with frost damage. Dead stems are often scattered in the field with straw-colored to pearly white dead shoots. Infected crowns turn blue-black, produce fewer stems per plant, and the plant eventually dies. Moderate or higher resistance is available in many varieties.

A new race of this disease has been identified in the Midwest that

Distribution and severity
Anthracnose

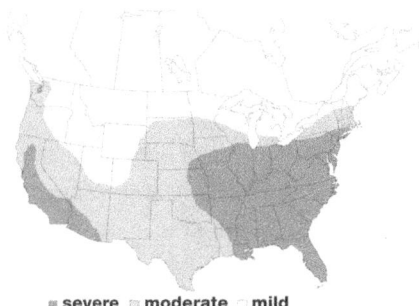

■ **severe**　□ **moderate**　□ **mild**

overcomes the resistance. If planting a HR variety to race 2 and still seeing the disease, it may be worthwhile to use a foliar fungicide to control this disease during the growing season.

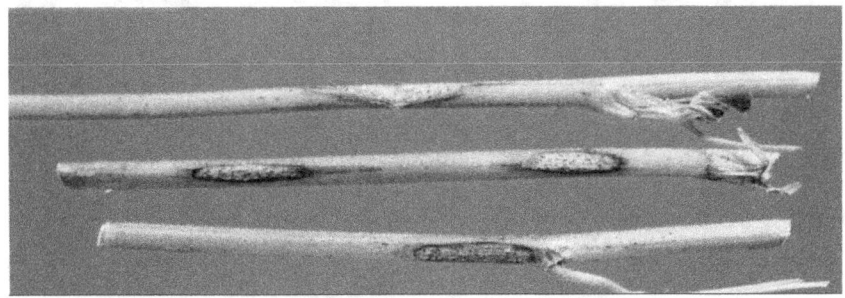

Straw-colored lesions on stems are indicative of anthracnose.

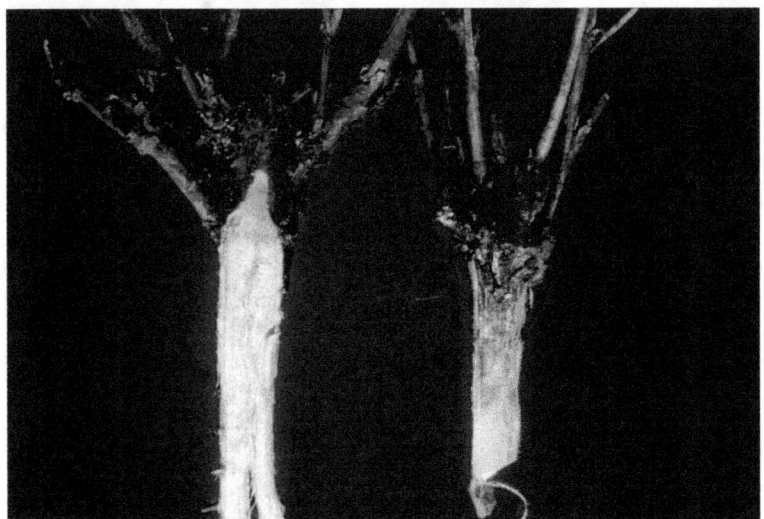

The diseased crown (right) shows blue-black coloring of anthracnose.

Aphanomyces root rot

Aphanomyces root rot is an important
disease of wet soils. It stunts and kills
seedlings and causes a chronic root
disease in established plants. Infected
seedlings develop yellow cotyledons
followed by chlorosis of other leaflets.
Roots and stems initially appear gray
and water-soaked, then turn light to
dark brown. Seedlings become stunted
but remain upright. Aphanomyces
reduces root mass on established plants.
Nodules are frequently absent or in
some stage of decay. Infected plants
exhibit symptoms similar to nitrogen
deficiency and are slow to regrow
following winter dormancy or harvest.
For best results, select varieties with
high levels of resistance to both aphanomyces and Phytophthora root rot. There
are two races of aphanomyces. Race 1 is
the most common form; however, Race 2
occurs in many areas and is more viru-
lent than Race 1. If you plant a resistant variety and still have the disease,
select a variety with Race 2 resistance.
Additionally, areas with severe aphanomyces, will benefit from planting
seed additionally treated with both the
fungicides Apron and Stamina.

Comparison of susceptible (left) and resistant (right) varieties shows
stunting and slight yellowing caused by aphanomyces.

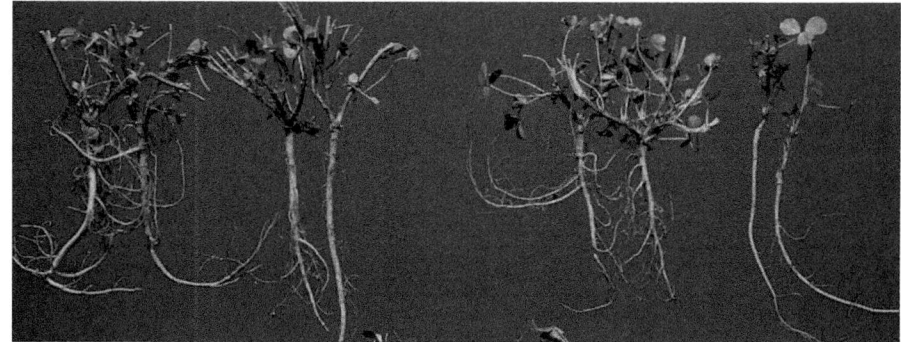

Infected plants lack lateral roots (from left, second and fourth pairs).

Distribution and severity
Aphanomyces

■ **severe** ▪ **moderate** ▫ **mild**

Infected seedlings develop yellow cotyledons.

Bacterial wilt

Bacterial wilt symptoms begin to appear in the second and third year and may cause serious stand losses in 3- to 5-year-old stands. In early stages, affected plants turn yellow-green and are scattered throughout the stand. Severely infected plants are stunted with many spindly stems and small, distorted leaves. Diseased plants are most evident in regrowth after clipping. Cross sections of the taproot show a ring of yellowish brown discoloration near the outer edge. Most varieties are now resistant to this disease.

The entire plant is stunted and yellowed by bacterial wilt.

Distribution and severity
Bacterial wilt

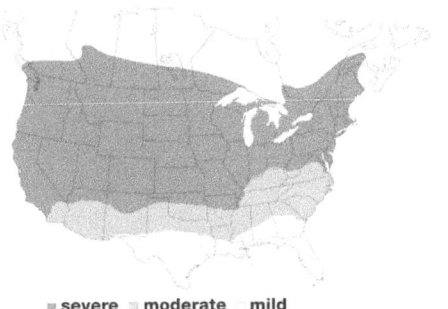

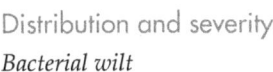
■ **severe** ■ **moderate** ■ **mild**

Varying degrees of infection shown by yellowish-brown ring.

Common leaf spot and lepto leaf spot

Common leaf spot occurs primarily in first and second cuttings and in fall regrowth of most alfalfa stands. Disease severity depends on alfalfa conditions and varietal resistance. Symptoms appear as small, brown to black lesions—each less than 0.1 inch diameter—that rarely grow together. On the upper leaf surface, the lesions may have a small raised disc in the center. Leaves turn yellow and fall off. The disease causes yield reductions and lowered forage quality through leaf loss. Severely infected fields should be harvested early. Some varieties are moderately resistant.

Lepto leaf spot attacks young regrowth of alfalfa during spring and fall or midwinter in southern areas. Disease growth is particularly noticeable following cool, rainy periods. The lesions start as small, black spots and enlarge to 0.1 inch in diameter with light brown or tan centers. The lesions are usually surrounded by a yellow, chlorotic area. Lesions often enlarge and grow together. Yield and quality is lost through loss of dead leaves by wind or during harvesting. Resistant cultivars are not available.

Distribution and severity
Common leaf spot

■ severe ■ moderate ▢ mild

Distribution and severity
Lepto leaf spot

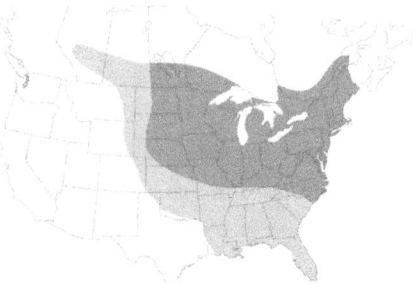

■ severe ■ moderate ▢ mild

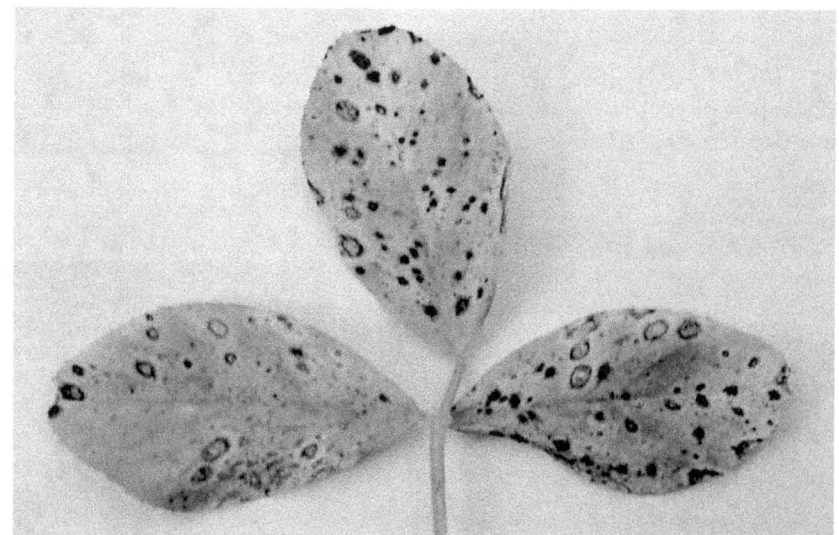

Lepto leaf spot lesions have tan centers and are surrounded by a yellow halo. Lesions often enlarge and grow together.

Common leaf spot lesions are small brown to black areas that rarely grow together.

Fusarium wilt

Fusarium wilt is a vascular disease that causes gradual stand thinning. Initially, plants wilt and appear to recover overnight. As the disease progresses, leaves turn yellow then become bleached, often with a reddish tint on only one side of a plant. After several months the entire plant dies. Symptoms are similar to bacterial wilt but plants are not stunted. To diagnose Fusarium, cut a cross section of the root. The outer ring (stele) of the root is initially streaked a characteristic reddish-brown or brick red color. As the disease progresses the discoloration encircles the root and the plant dies. Practice good fertility and control pea aphids and potato leafhoppers to reduce the effects of this disease. Many varieties are resistant to Fusarium wilt.

Distribution and severity
Fusarium wilt

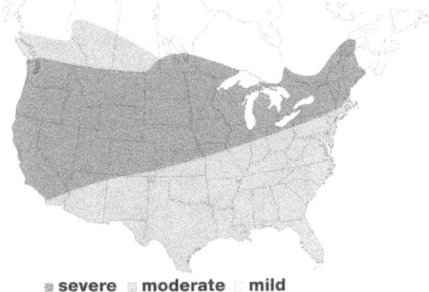

▪ **severe** ▪ **moderate** ▪ **mild**

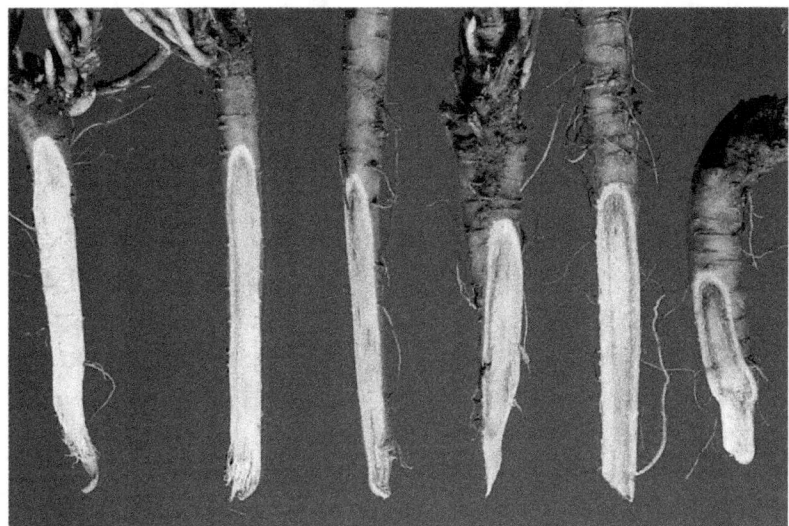

Fusarium causes a characteristic reddish-brown discoloration that becomes more evident as the disease progresses (left to right).

Disease bleaches the leaves and stems on plants scattered throughout the field. Symptoms are similar to bacterial wilt, but affected plants are not stunted.

Phytophthora root rot

Phytophthora root rot can kill seedlings and established plants in wet or slowly drained soils. The disease is especially prevalent among new seedlings in cool, wet soils. Infection occurs as plants emerge; they appear water-soaked and then collapse and wither. The disease appears on established plants in poorly drained soils and where water stands for 3 days or less. Plants wilt; then leaves, especially lower ones, turn yellow to reddish brown. Lesions develop on the roots. In severe cases, taproots may rot off at the depth of soil water saturation (frequently 1 to 6 inches below ground surface). Plants may die within 1 week of infection or linger on with reduced root mass and growth rate. Often Phytophthora root rot is not discovered until the soil dries and apparently healthy plants begin wilting because their rotted taproots are unable to supply adequate water. Many highly resistant varieties are available for poorly drained soils.

Crop rotation is of little value for Phytophthora root rot control because the fungus can survive indefinitely in the soil. However, good management practices can prolong the productivity and life of infected plants that survive the initial infection.

1. Maintain high soil fertility to promote extensive lateral root development above the diseased region of the root and to extend the life of the plant.

2. Avoid untimely cuttings that might stress the plants. Heavy rains immediately after cutting often result in severe infections. Do not cut, for example, between September 1 and October 15.

3. Control leaf-feeding insects, which can stress plants and make them more susceptible to Phytophthora.

4. Tilling and land-leveling, if practical, can reduce Phytophthora root rot by improving surface and subsurface drainage.

Distribution and severity
Phytophthora root rot

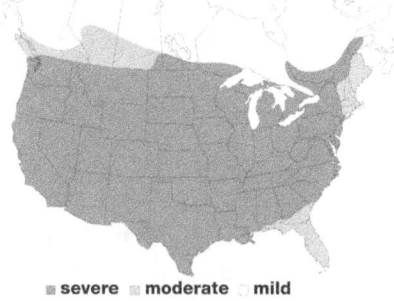

■ **severe** ■ **moderate** □ **mild**

Stems and leaves are bleached by Phytophthora.

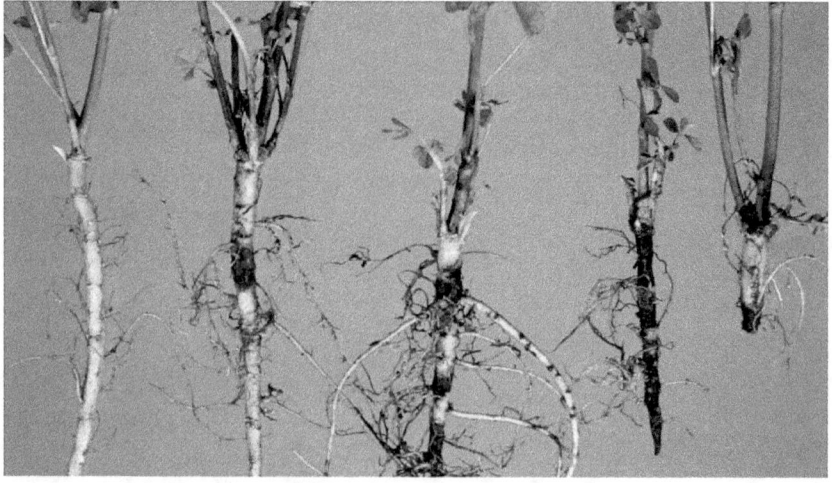

As the disease progresses (left to right), lesions develop and the taproot rots off.

Root-lesion nematodes

Root-lesion nematodes reduce yield and thin stands. The parasitic nematodes are microscopic worms that feed on root hairs, feeder roots, and nitrogen-fixing nodules of alfalfa. Root-lesion nematodes reduce the alfalfa plant's ability to take up soil nutrients and fix nitrogen. Plants appear unhealthy and stunted, usually in spotty areas within an otherwise healthy stand. Nematode populations can be reduced by rotating to row crops or fallowing for 2 months following incorporation of forage crop residue. Moderate resistance is available in some varieties.

Distribution and severity
Root-lesion nematodes

■ **severe** ■ **moderate** □ **mild**

Stunted plants and stand thinning caused by root-lesion nematodes.

Seedling death due to root-lesion nematodes.

Sclerotinia

Sclerotinia crown and stem rot is most damaging to seedling stands, especially those seeded in late summer. The first symptoms appear in the fall as small, brown spots on leaves and stems. During the cool, wet weather of early spring, the crown or lower parts of individual stems soften, discolor, and disintegrate. As infected parts die, a white, fluffy mass grows over the area and hard, black bodies, known as sclerotia, form. These bodies remain on the surface of the stem or become imbedded in it. Infection will spread if cool, wet weather prevails during spring, causing rapid thinning of stands. Spring planting reduces incidence of the disease. Plowing buries sclerotia and reduces its ability to infect new plantings. Some resistance is available in some varieties.

Distribution and severity
Sclerotinia

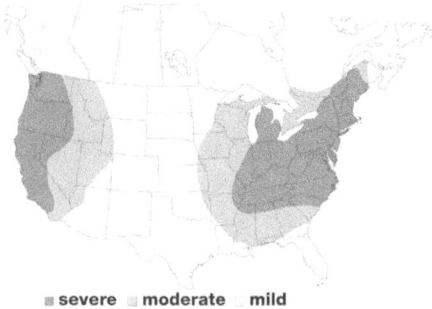

■ **severe** ■ **moderate** ▪ **mild**

Arrows point to white fluffy masses at the base of a stem.

Softened and discolored stems.

Sclerotia on stem.

Spring black stem

Spring black stem occurs in the northern United States during early spring and reduces forage yield and quality. Many small, dark brown spots develop on the lower leaves and stems. Leaves, especially lower ones, turn yellow, wither, and fall off. Lesions on stems enlarge and may blacken large areas near the base of the plant. Severe infestations girdle and kill the stem. The plant dies when infection spreads to the crown and roots. Cutting the stand at early stages of maturity will reduce leaf loss and disease prevalence. Currently available varieties have variable levels of resistance, but none are characterized for this disease.

Distribution and severity
Spring black stem

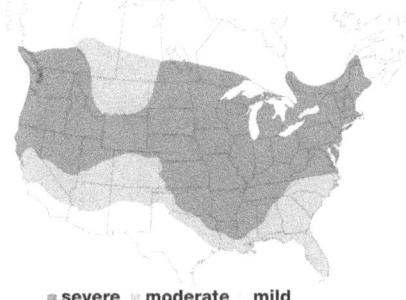

■ **severe** ■ **moderate** ▫ **mild**

Summer black stem

Summer black stem occurs during hot, humid weather, reducing forage yield and quality. The disease first affects the base of the plant and progresses up the stem, causing leaves to fall off. Leaf spots are brown with irregular margins and often surrounded by a diffuse yellow margin. Reddish to chocolate brown oval lesions form on the stems and merge to discolor most of it. Early harvest may reduce losses. Currently available varieties have little resistance.

Distribution and severity
Summer black stem

■ **severe** ■ **moderate** ▫ **mild**

Lesions may enlarge to girdle stems and kill the plant.

Leaf lesions (left) first appear on lower leaves.

Verticillium wilt

Verticillium wilt can reduce yields up to 50% beginning the second harvest year and severely shortens stand life. Early symptoms include v-shaped yellowing on leaflet tips, sometimes with leaflets rolling along their length. The disease progresses until all leaves are dead on a green stem. Initially, not all stems of a plant are affected. The disease slowly invades the crown and the plant dies over a period of months. Root vascular tissues may or may not show internal browning. Many varieties are resistant to this disease.

The following measures minimize the chances of introducing the fungus to an area and spreading the disease between and within fields.

1. Plant resistant varieties.

2. Practice crop rotation. Deep plow Verticillium-infested fields and do not plant alfalfa for 2 to 3 years, although a highly resistant variety could be planted sooner. Corn and small grains are important non-hosts. These crops should fit well into a rotation with alfalfa. Red clover is a questionable host, so don't grow red clover on Verticillium-infested land.

3. Harvest non-infested fields first. Then harvest infested fields at the hard-bud or early flower stage. Early harvest can limit some yield and quality losses caused by Verticillium wilt and can slow the spread of the fungus in a field.

Browning in roots.

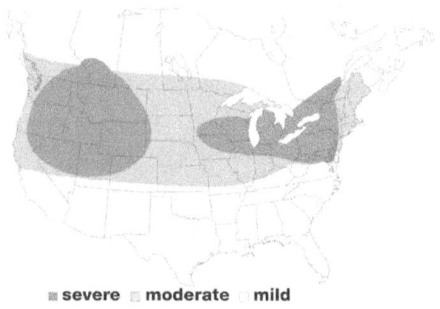

Distribution and severity
Verticillium wilt

■ severe   ▨ moderate   ▢ mild

At later stages of infection, dead leaves appear on green stems.

Leaves show early symptoms of Verticillium wilt. Note v-shaped yellowing and scattered bleaching.

Insect management

Alfalfa blotch leafminer

The alfalfa blotch leafminer was first detected in the Midwest in 1996. Adults are small, black, hump-backed flies that emerge from overwintering pupae located on the surface of the soil. The first indication of their presence is the appearance of numerous pinholes (from a few to over 100) in the alfalfa leaflets. These pinholes are punctures made during egg laying, but the adults also feed on plant material that oozes from the punctures. Females lay one to three eggs per leaflet. Small yellow maggots hatch within the leaf and begin feeding between the upper and lower leaf surfaces. As the leafminers eat their way from the base of the leaflet toward the tip, the tunnel, or mine, they create widens as they mature. The resulting tunnels give the leaflets a blotchy appearance. When fully grown, the leafminers crawl out of the leaves, drop to the ground, and pupate. In the upper Midwest, a second generation of flies emerge in mid-July, and a third generation follows in late August.

Damaged leaves have reduced protein content and may fall off. Significant yield loss should only occur if damaged leaves drop or are shaken from the forage during harvest.

In the upper Midwest, harvest of the first crop normally controls the first generation. Development of the second and third generations, however, may not correspond as closely with cutting schedules and this could lead to more extensive injury in those cuttings. Insecticidal control may be warranted if at least 30% of the leaflets have pinhole injury. Delaying application until blotches are apparent on numerous plants will reduce insecticide effectiveness. Because the eggs hatch over an extended period and the adults are mobile, some insecticide trials have had marginal control results. Biological control of this pest is well established in the northeastern United States. It is anticipated that biological control will also be a major control factor in the Midwest as parasitized larvae were detected in Wisconsin in 1998.

Alfalfa weevil

Alfalfa weevil larvae chew and skeletonize leaves. Large larval populations may defoliate entire plants, giving the field a grayish cast. Damage normally only occurs to the first harvest but both larvae and adults may damage regrowth when populations are high, resulting in both yield and stand loss.

Larvae are slate-colored when small, but bright green when full grown (3/8 inch). They have a white stripe down the back and a black head. Although larvae are present from May well into the summer, peak feeding activity falls off by mid-June.

When full grown, the larvae spin silken cocoons on the plants, within the curl of fallen dead leaves, or within litter on the ground. Adults emerge in 1 to

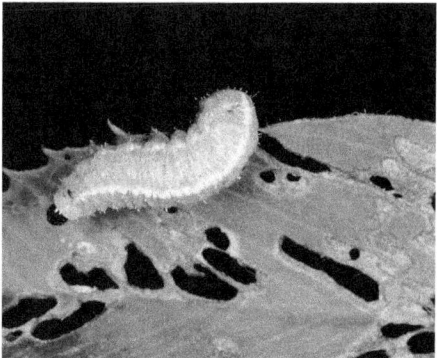

Alfalfa weevil larva and feeding damage.

Distribution and severity
Alfalfa blotch leafminer

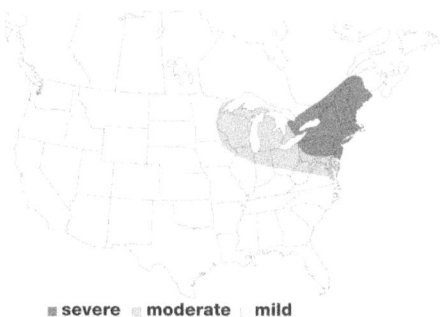

■ severe ■ moderate ▪ mild

Tunneling damage caused by alfalfa blotch leafminer larvae.

Plants severely damaged by alfalfa weevil feeding appear grayish-brown.

2 weeks. They are dark gray to brown snout beetles measuring 3/16 inch in length. There is a distinct dark shield-like mark on the back. After feeding a short time, most leave the field and enter a resting period that lasts until fall. In the fall, they return to the alfalfa field and lay a few eggs before the onset of cold temperatures. In northernmost states, fall egg laying is insignificant; most eggs are laid the following spring.

Begin checking alfalfa fields for signs of weevil feeding around mid-May in northernmost states and earlier farther south.

Treat fields when larval counts average 1.5 to 2 per stem or 40% of the plant tips of the first crop show obvious signs of damage. *This does not mean 40% defoliation.* If damage occurs within 7 to 10

days of the suggested harvest date, harvest the hay as soon as possible; otherwise spray the field as soon as possible. Many weevil larvae are killed during harvesting.

If you've harvested early because of developing alfalfa weevil problems, or if substantial weevil damage has occurred, check the stubble carefully for signs of damage to new growth. Some fields may fail to green-up because adults and larvae consume new crown buds as fast as they are formed. Examine the stubble, the soil surface

around alfalfa plants, and under leaf litter for larvae and adults. Stubble protection is rarely needed, but if there are more than eight larvae and new adults per square foot in the stubble or more than 50% of the new growth has been damaged, then spray the stubble as soon as possible.

If most damage to regrowth is being caused by adults, check the insecticide label to make sure the product is registered for adult control and that a high enough rate is applied.

Distribution and severity
Alfalfa weevil

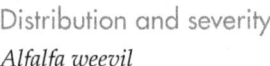
■ severe ▨ moderate ▢ mild

Adult weevils bear a distinctive shield-like mark on the back. Larvae are slate-colored when small but turn bright-green when full grown.

Aphids

Aphids cause stunting and yellowing of alfalfa resulting in yield loss. Heavily infested plants wilt during the hottest parts of the days.

Green pea aphids (pictured), spotted alfalfa aphids (yellow with faint dark spots), or cowpea aphids (with velvety black appearance and distinct waxy cover) congregate on stems and leaves and suck plant juice. Spotted alfalfa aphids have been uncommon in the upper Midwest for many years. Parasites and disease keep the pea aphid in check most years, though population explosions periodically occur. Pea aphids are a major problem in the hot and dry western United States. Treat pea aphids when numbers exceed 100 per sweep, particularly during dry periods.

Distribution and severity
Pea aphid

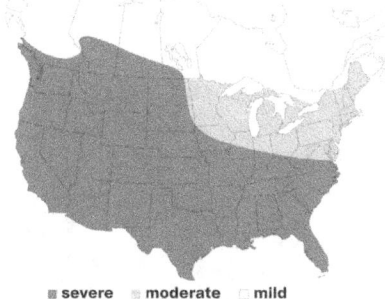

■ severe   ■ moderate   ☐ mild

Distribution and severity
Blister beetles

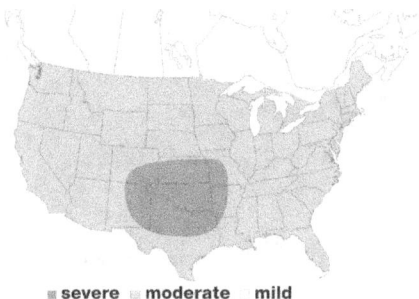

■ severe   ■ moderate   ☐ mild

Blister beetles

Blister beetles in alfalfa hay can cause sickness and death in livestock, particularly horses. Blister beetles contain cantharidin, a chemical irritant that can blister internal and external body tissues. Although there are few documented cases of fatalities in cattle and sheep, cantharidin-contaminated hay is deadly to horses. The amount of cantharidin necessary to kill a horse is 1 milligram per kilogram of horse weight. Blister beetles vary in toxicity depending on the species. It would require 100 striped blister beetles to kill a 1200 lb horse compared to 1100 of the less toxic black blister beetles.

Blister beetles are a serious problem in southern and western states, and an occasional problem in the upper Midwest, particularly during drought years or the year following drought. The several species present in the Midwest vary in size and color, but are easily recognized by their elongated, narrow, cylindrical, soft bodies. The "neck" area is narrower than on most beetles. Scouting is misleading because the beetles tend to cluster and will be concentrated in parts of the field

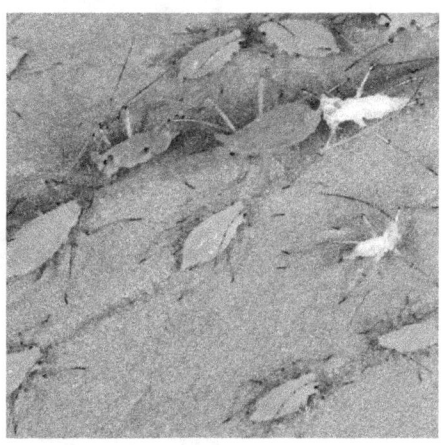

Aphids congregate on leaves and stems to suck plant juice

Black blister beetle.

Striped blister beetle.

Margined blister beetle.

Ash-gray blister beetle.

while absent from other parts. Sprays are generally not effective because cantharidin is a very stable compound and the dead beetles can be picked up in the hay. Because beetle populations tend to build throughout the season, especially in the south, horse owners should consider buying first-crop and early second-crop hay during high infestations of blister beetle. Harvesting fields prior to flowering and maintaining weed-free stands will reduce beetle populations.

Clover leaf weevil

Clover leaf weevil larvae eat alfalfa leaves, usually beginning with the foliage around the base of the plant. Crop injury occurs mostly before the first cutting, but it is usually insignificant compared with the injury caused by the alfalfa weevil. Clover leaf weevils are active at night and on cloudy days. During sunny days, they hide around the base of the plant.

Larvae are slate-colored when small, and bright green when full grown. They are similar in appearance to alfalfa weevil larvae except that the head is light brown and the white stripe down the center of the back is often edged with pink. Full-grown larvae are about 1/2 inch long.

Distribution and severity
Clover leaf weevil

■ **severe** ■ **moderate** ■ mild

Adult clover leaf weevils are two to three times larger than alfalfa weevils. They are 5/8 inch long, dark brown flecked with black, and have a lighter colored stripe extending along each side of the wing covers. This insect normally leaves the fields shortly after the first cutting and returns in late summer to feed and lay eggs before winter. There is one generation per year and they overwinter mostly as partially grown larvae.

Treatment is rarely warranted for clover leaf weevil larvae. Management of alfalfa weevils will also control this insect. However, adult clover leaf weevils can cause damage by feeding on the green stems and regrowth after the first cutting. Large populations can cause extensive feeding damage, scarring the stems and rapidly consuming new foliage as it is produced. This type of injury is more common during dry springs when regrowth is slow and weevils are abundant. Treatment should be considered if plants do not begin to regrow in 3 to 4 days after cutting and weevils are present in the field.

Adult clover leaf weevils are dark brown flecked with black.

Clover leaf weevil larvae hide around the base of the plant during sunny days, preferring to feed at night and on cloudy days.

Clover leaf weevil larvae are similar to alfalfa weevil larvae but have light brown heads rather than black.

Clover root curculio

The clover root curculio is a potentially serious pest of alfalfa. Although this pest can be found in most alfalfa fields, high populations and serious damage have been localized and sporadic. However, even small populations may contribute to stand decline. At this time there is no reliable method of damage prediction or control.

Adults are black to dark brown, blunt-snouted weevils that are approximately 1/8 inch long and 1/16 inch wide. The surface of the beetle's body is deeply "punctured." Females lay eggs on the lower parts of stems, on lower leaves, or on the soil surface. Larvae hatch from these eggs and enter the soil through surface cracks.

There is only one generation per year. Adults lay eggs in fall or spring, and hibernate over the winter. Eggs hatch in the spring, and egg-laying is usually complete by mid-June. New adults emerge in June and July and live about a year.

Adult curculios injure plants by chewing the margins of leaves, creating crescent-shaped notches, or by chewing the stems and leaf buds of young seedlings. Feeding damage can weaken seedlings, causing poor growth or death. Mature plants are not at risk unless populations are exceedingly high.

Larvae do the greatest damage, and such damage can be cumulative over the years that a field exists. Newly hatched larvae feed on nodules and small rootlets and chew out portions of the main root. Feeding on the main root leaves long brown furrows and may partially girdle the plant.

Damage from clover root curculios is believed to shorten stand life, contribute to winter kill, and provide an avenue for entry by disease organisms. No commercially acceptable control techniques are available. However, do not plant alfalfa back into old, infested alfalfa stands because curculio damage can destroy the new stand. Also, since adults migrate primarily by crawling from field to field, avoid seeding alfalfa next to older stands.

Long brown furrows on the taproot caused by curculio larvae feeding.

Distribution and severity
Clover root curculio

■ **severe** ■ **moderate** ▫ **mild**

Grasshoppers

Grasshoppers can overwinter as eggs or adults, depending upon the species. Populations tend to build during the season, followed by movement of the grasshoppers into cultivated crops from grassy or weedy areas where they overwintered. It is important to detect infestations while the grasshoppers are small and concentrated in overwintering sites.

Several species can feed on alfalfa. Problems occur mainly in the western United States and during droughty years in the Midwest. Grasshoppers rarely cause economic damage in most areas of the Midwest and should be considered a minor pest.

Begin spot-checking overwintering sites during June. Estimate the number of grasshoppers per square yard while walking through these areas. Insecticide use is not suggested until populations reach 20 per square yard in field margins or 8 per square yard within an alfalfa field. If economically damaging infestations are detected while the grasshoppers are still concentrated, spot treat the area to protect alfalfa fields.

Plant bugs

Plant bugs extract plant sap with their tube-like mouths. High populations can stunt alfalfa growth or crinkle and pucker leaves. However, these symptoms may be caused by other factors so be sure to positively identify the problem before treating plants.

The two plant bugs that are particularly important to alfalfa production are the tarnished plant bug and the alfalfa plant bug. The adult tarnished plant bug is 1/4 inch long and brown. Nymphs are green with black spots on the back. Adult alfalfa plant bugs are 3/8 inch long and are light green. Nymphs are green with red eyes.

Treatment is suggested if there are three plant bug adults and/or nymphs per sweep on alfalfa that is less than 3 inches tall; treat when there are five or more adults and/or nymphs per sweep on taller alfalfa. If damage occurs within 7 to 10 days of the suggested harvest date, harvest the hay as soon as possible; otherwise spray the field as soon as possible.

Distribution and severity
Grasshoppers

■ **severe** ■ **moderate** ■ **mild**

Grasshoppers are best controlled when they are in field borders and before they move into the alfalfa.

Adult tarnished plant bugs are 1/4 inch long.

Alfalfa plant bugs are 3/8 inch long.

Distribution and severity
Plant bugs

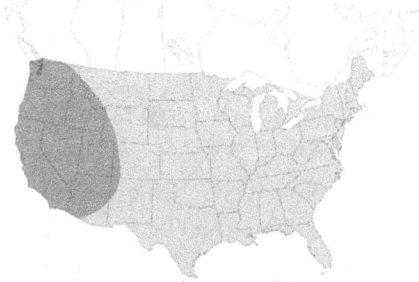

■ **severe** ■ **moderate** ■ **mild**

Crinkled leaves typical of plant bug damage.

Potato leafhoppers

Potato leafhoppers are mid- to late-season alfalfa pests that migrate to northcentral and eastern states from southern areas in late spring. First-crop alfalfa harvested at the proper time in the Midwest usually escapes damage. However, subsequent crops and new seedlings should be monitored for leafhoppers.

These small (1/8 inch), green, wedge-shaped insects suck sap from plants and damage the phloem of leaves, restricting water and nutrient flow to the outer tip of the leaf. This creates a yellow wedge-shaped area on the tip of leaflets. Severely damaged plants will be stunted, and chlorosis will appear on all leaves if leafhoppers are not controlled. Damage first appears along the edges of fields.

Alfalfa stands suffer yield and quality losses before any yellowing is visible. To detect leafhoppers before symptoms appear, scout fields using an insect sweep net. Count adult and nymph leafhoppers in 10 sweeps covering several areas of the field. The decision to spray depends on the following factors:

1. Whether alfalfa is a new seeding or established stand. New seedings are most susceptible; damage in the first year can reduce yield for the life of the stand. It is particularly important to control potato leafhoppers on new seedings under a cover crop by either scouting and spraying or using a resistant variety, otherwise stands may die out.

2. Plant height. Taller plants are able to tolerate more leafhoppers.

3. Whether or not the variety has greater than 50% resistance to potato leafhopper. Leafhopper population growth is inhibited in highly resistant varieties. Resistant varieties suffer significantly less damage and require insecticide treatment less frequently than susceptible varieties.

Refer to Figure 16 to determine appropriate action to take in alfalfa fields. These spray guidelines are based on average costs of insecticide treatment and average hay value. Growers should consider altering the action thresholds if treatment cost or hay value deviates greatly from average.

Distribution and severity
Potato leafhoppers

■ **severe** ■ **moderate** □ **mild**

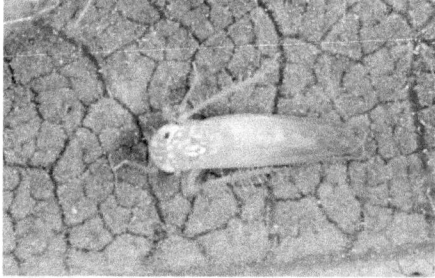

Adult leafhopper (actual size 1/8 inch).

Figure 16. Economic action thresholds for control of potato leafhopper (PLH) in alfalfa.

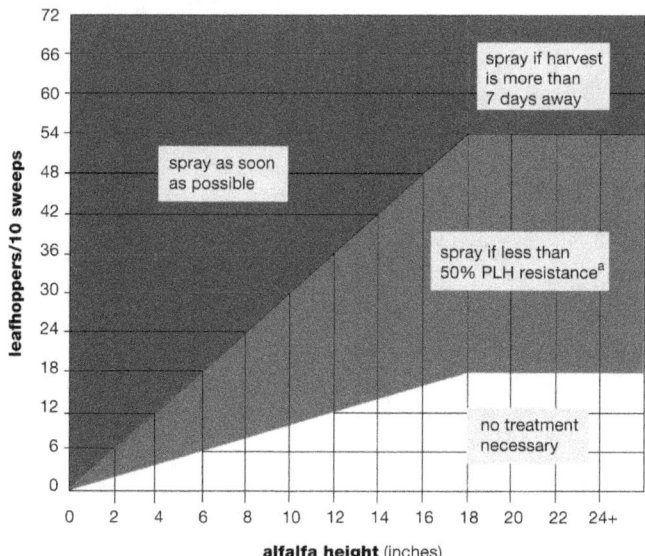

Source: Mark Sulc and Ron Hammond, The Ohio State University, 2004.

Severely damaged plants are stunted and chlorotic. Leafhopper burn appears first as yellow wedge-shaped areas on the tips of leaflets.

Spittlebugs

Spittlebug nymphs appear in early May. These soft, orange or green bugs can be found in white spittle masses in leaf axils, and later in the clumps of new growth at tips of stems. They suck plant juices and stunt but do not yellow the alfalfa. Alfalfa can support a tremendous population of spittlebugs without yield loss and they usually have no economic impact. Treatment is suggested if there is an average of one spittlebug per alfalfa stem.

Distribution and severity
Spittlebugs

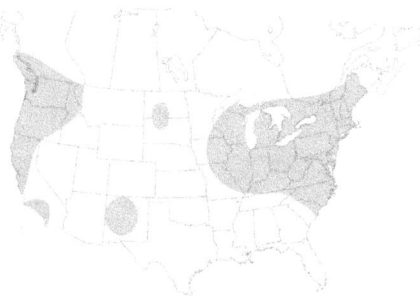

■ **severe** ■ **moderate** ■ **mild**

Spittlebug froth.

Variegated cutworm

Variegated cutworm larvae feed on leaves and stems. Serious damage can occur on regrowth after the alfalfa is cut and larvae feed under the protection of drying windrows. They also can cut seedling plants in new stands. Larvae are variable in color, ranging from tan to greenish-yellow to almost black with a row of small yellow, dagger- or diamond-shaped spots down the center of the back. There are three to four generations a year.

Treatment should be considered if the hay does not begin to regrow in 4 to 7 days after cutting and larvae are present in the field.

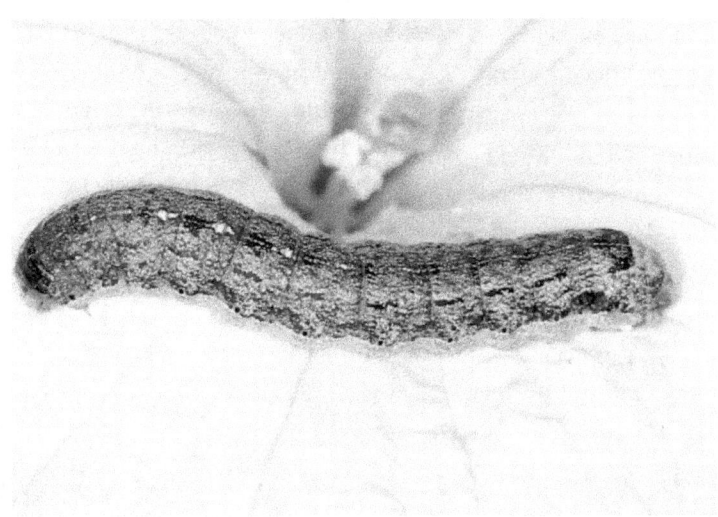

Variegated cutworm larvae can cause serious damage on re-growth after alfalfa is cut.

Distribution and severity
Variegated cutworm

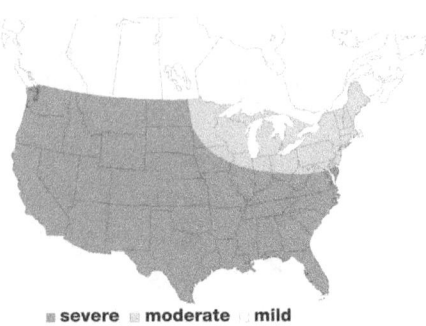

■ **severe** ■ **moderate** ■ **mild**

When to rotate from alfalfa

To decide when to rotate from alfalfa, you'll need to evaluate stand density and yield relative to your needs. You'll also want to factor in rotation requirements, farm plan, total acreage of forage needed, and ability to reseed. Because most of these factors are farm specific, this section focuses on the relationship between stand density and yield.

Alfalfa has a tremendous ability to produce maximum yield over a wide range of stand densities. New seedings should have at least 25 to 30 plants per square foot the seeding year. Stands gradually thin and weeds may invade rapidly. Weedy stands force the choice of using herbicides, which increases production cost, or of harvesting much lower quality forage.

The decision to reseed new fields of alfalfa should be based on the yield potential of the stand, ideally using actual yields from the field. The next best method is to count stems when the alfalfa is 4 to 6 inches tall and use the data from Figure 17 to estimate yield potential (assuming drought, soil fertility, or other conditions are not limiting yield). In the Midwest, the Northeast, and many irrigated fields in other regions, yields often begin to decline in the third year of production. Fields with reduced yields still cost about the same as high-yielding fields. This is because high-yielding fields require less herbicide to produce high-quality forage. Plowing down more dense stands will produce nitrogen credits. There is also a rotational benefit to corn following alfalfa: it yields 10 to 15% more than corn following corn.

The best time to make stand decisions is in the fall. During the last growth period record stem density. Then dig a random sampling of plants and assess root health (see related advanced technique). Typically, stands that fall below 40 stems per square foot or three to four

Advanced Techniques

Stand evaluation

To evaluate stands, dig several alfalfa plants in the fall and look at the condition of the root. This will give an idea of stand vigor and future life span. Some crown rot will be visible in most older stands. Look for the number of crowns and roots with rot and the degree of infection.

Categorize plants using a scale of 0–5 (compare to the photographs on the following page). Determine the percentage of plants in each category. Healthy stands have fewer than 30% of the plants in categories 3 and 4.

rating	winter survival
0	excellent
1	excellent
2	good
3	marginal to severe winter kill
4	severe winter kill
5	already dead

Individual plants with severe injury (greater than 50% rot) are not likely to survive another year. Stands with a high percentage of these plants should be considered for replacement. Thus you can use stem count to determine yield potential now and the plant root assessment to determine whether the yield will be the same or less next year. Based on this you can decide whether or not to keep the stand.

Figure 17. Alfalfa stem count and yield potential.

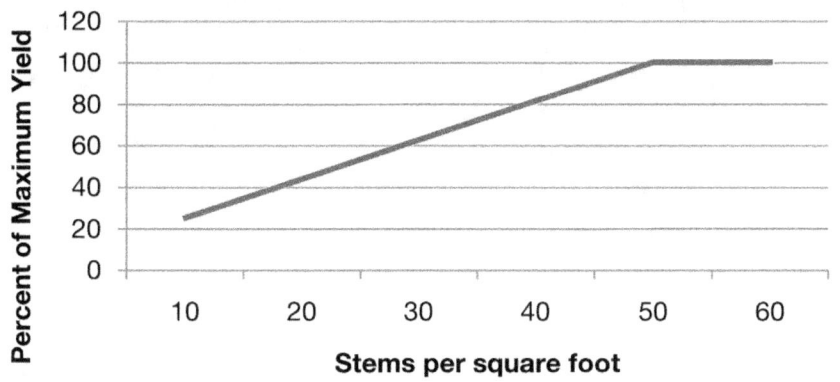

Source: Undersander and Cosgrove, University of Wisconsin, 1992.

healthy plants per square foot are no longer profitable, although the critical yield range will vary with individual farming operations. Marginal stands that are healthy may be kept while fields with high levels of crown rot will decline rapidly and should be considered for rotation along with low yield potential fields.

Varying degrees of crown rot from upper left, healthy roots, to lower right, severe crown rot resulting in death.

Healthy plant

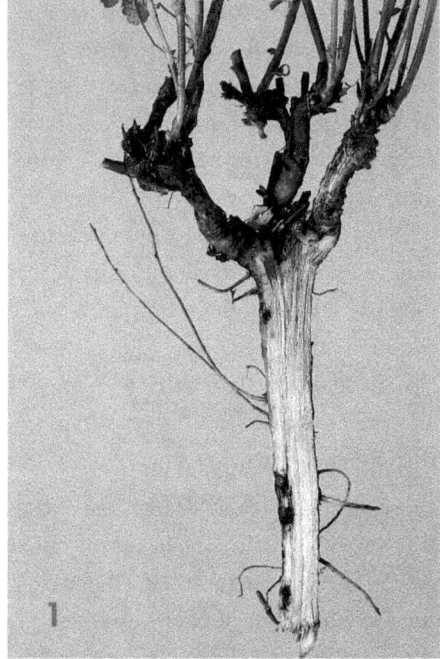

Some discoloration

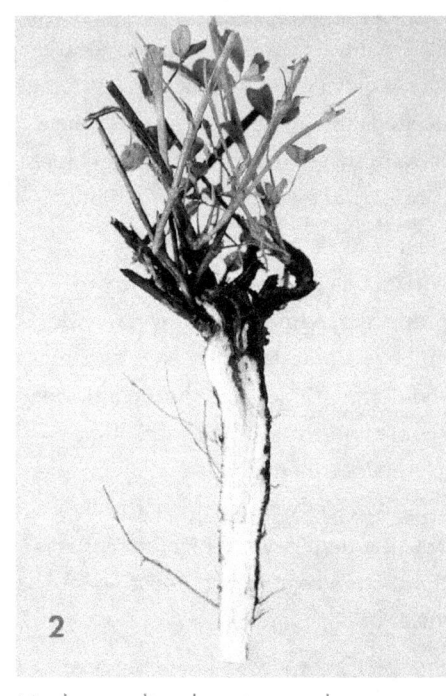

Moderate discoloration and rot

Significant discoloration and rot

Greater than 50% discoloration

Dead

Harvest

The final step to profitable alfalfa production is to set goals for forage quality and use the appropriate harvest techniques to minimize field losses and maximize tonnage of high quality forage. This recognizes that high quality forage is profitable to animals that can use the quality but that tradeoffs exist between forage quality, yield, and stand life.

Forage quality

Alfalfa is superior to other forage crops because it is high in crude protein and energy, reducing the need for both types of supplements in rations. The superior intake potential allows for greater use in rations of high-producing dairy cows.

What quality forage is needed?

The nutrient need of an animal depends primarily on its age, sex, and production status (Figure 18). Maximum profit results from matching forage quality to animal needs. Lower-than-optimum quality results either in reduced animal performance or increased supplement costs. Conversely, feeding animals higher quality forage than they need wastes unused nutrients that were expensive to produce and may result in animal health problems.

Quality standards are presented in Table 10. (Forage quality terms are defined at the end of the Harvest section.) Use the RFQ index to allocate the proper forage to the proper livestock class (Figure 18). Performance of high-producing dairy cows is most limited by intake of digestible dry matter and prime hay or haylage is recommended. An RFV or RFQ of 151 or higher is recommended for dairy cows after the first trimester, heifers, and stocker cattle.

As shown in Figure 19, ADF is a poor estimate of energy in feedstuffs. In response, the National Research Council Nutrient Requirements for Dairy Cattle (2001) recommended estimating energy from total digestible nutrients (TDN). TDN is the sum of digestible components (nonfibrous carbohydrates, crude protein, fatty acids, and digestible fiber; see page 59).

Table 10. Quality standards for legume, grass, and grass–legume mixture.

RFQ†	NDF
>151	<40
151–125	40–46
124–103	47–53
102–87	54–60

Abbreviations: RFQ = relative forage quality; NDF = neutral detergent fiber.
† Assumes a neutral detergent fiber digestibility (NDFD) value of 45 or higher.

Figure 18. Forage quality needs of cattle and horses.

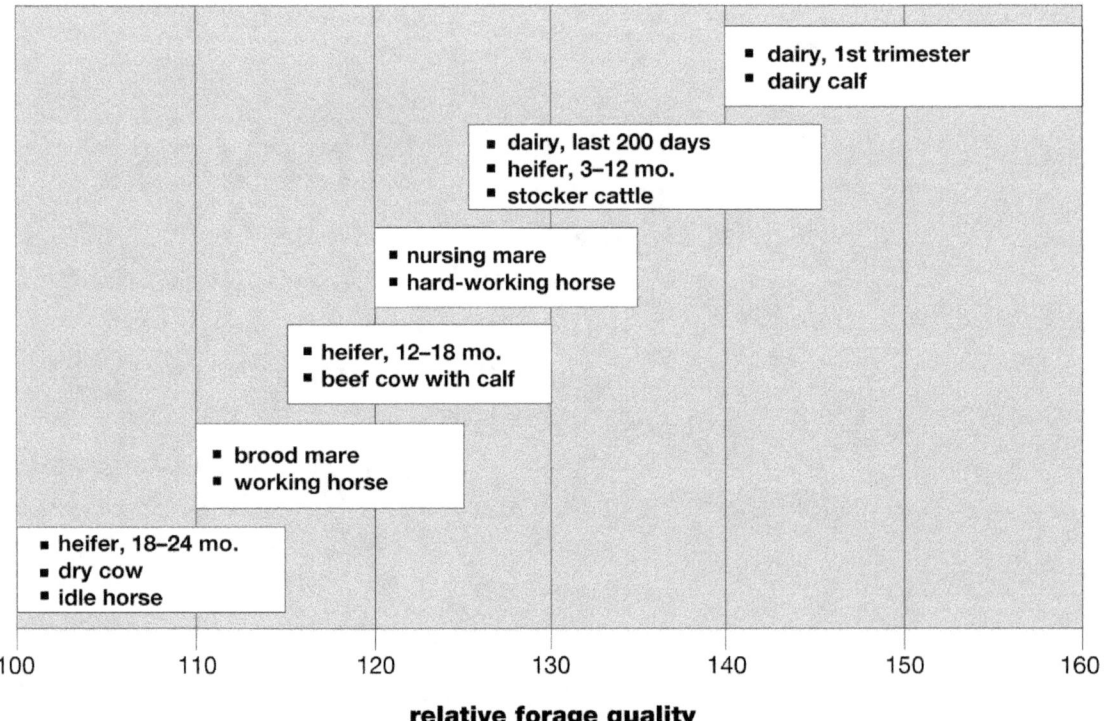

If NDFD is not reported on lab analyses, TDN was likely estimated from ADF only and is much less accurate.

RFQ was designed to use the new analyses that better predict animal performance. It is based on energy intake estimates relative to a standard just like RFV was. The only differences are that intake is adjusted for digestible fiber and that energy is calculated as TDN using digestible fiber. This calculation allows more meaningful comparisons between alfalfa, alfalfa–grass mixtures, and grasses.

Plant growth and forage quality

Understanding how alfalfa grows and its relationship to forage yield, forage quality, and carbohydrate root reserves is critical to production of high quality hay. Alfalfa is a perennial plant that stores carbohydrates (sugars and starches) in the crown and root. Plants use these carbohydrate reserves for regrowth both in the spring and after each cutting. When alfalfa is 6 to 8 inches tall, it begins replacing carbohydrates in the root (Figure 20). This cycle is repeated after each cutting.

High levels of carbohydrate reserves encourage rapid regrowth after cutting and winter survival. Regrowth begins with buds either on the crown or at the base of old shoots (after first cutting). Alfalfa regrowth for second and later cuttings begins while growth from the previous cycle is beginning to flower. Cutting at late maturity can remove shoots for the next cutting and delay regrowth.

Forage growth is most rapid until early flowering (Figure 21). Forage growth continues until full flower, but often leaf losses from lower stems slow yield increase after first flower. Alfalfa forage quality is greatest in early vegetative stages when the leaf weight is greater than stem weight; however, by first flower, and sometimes earlier, stem proportion exceeds that of leaves. Higher alfalfa yields after early flower can be attributed mainly to more low-quality stems. As cutting interval increases or as plants are harvested at later stages of maturity, yield per cutting increases but quality of the forage harvested decreases.

Figure 19. Comparison of ADF to in vitro digestibility of alfalfa.

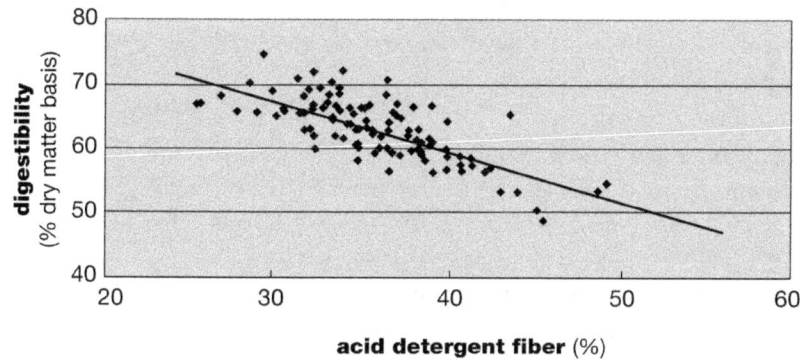

Figure 20. Carbohydrate content of alfalfa roots.

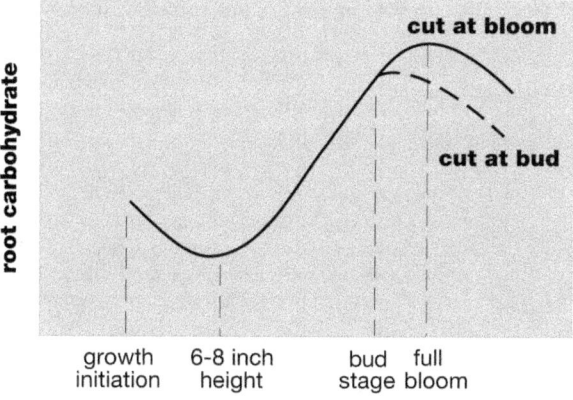

Figure 21. Forage yield relative to quality at different growth stages.

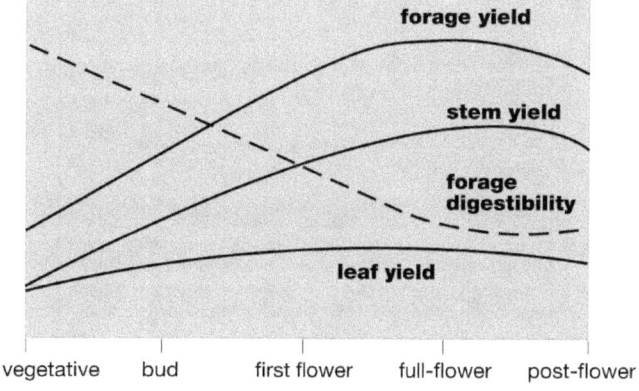

Temperatures during growth affect forage quality. Alfalfa grown during cool weather tends to produce higher quality forage than alfalfa grown during warm periods, assuming all harvests are equally weed-free and at the same maturity stage.

Forage quality is also influenced by the time of day alfalfa is cut. Plants convert sugars and starches to energy in a process called respiration. Respiration after cutting lowers forage quality and is stopped only by drying the forage. Therefore the best cutting time is early enough in the day to dry forage from the 75 to 80% when cut to less than 60% moisture by evening so that respiration is greatly slowed and starch and sugars preserved for animals. This generally means that morning cutting is recommended in the East either morning or after noon cutting is recommend in the more arid West.

Harvest management

Forage yield, quality, and stand persistence are all major considerations in the development of a profitable harvest management program. Increased awareness of the nutritional value of high quality alfalfa in terms of

potential savings of energy and protein supplements has caused many to re-evaluate current harvest strategies.

Cutting schedule

Selection of a harvest schedule begins with the decision on quality of forage desired. Growers desiring all high quality alfalfa will shorten stand persistence and decrease yield. Harvest schedule decisions include number of cuts per season, date of cut, stage of maturity, interval between cuts, and cutting height. The link between the stage of maturity and yield, quality, and persistence makes it apparent why growth stage is frequently used to decide when to harvest alfalfa. Keying harvests to specific stages of development also takes into account the varying effects of changing environments and variety maturity rates. A shortened growing season in northern states dictates combining calendar dates and stage of development into harvest strategies.

Maximum persistence

If harvesting for maximum persistence, cut alfalfa between first flower and 25% flower. This is approximately 35 to 40 days between cuttings (Figure 22). The system has a slightly wider harvest window and longer cutting interval

than when cutting for high quality because the emphasis is on high yield.

High quality

When harvesting for high quality the first cutting should be taken by an early calendar date appropriate for the region. The remainder of cuttings should be taken at midbud, generally 28- to 33-day intervals early in the season and longer near the end of the season (Figure 23). Cutting for high quality forage means that forage must be harvested within a 3- to 4-day period. No late-fall cutting should be taken in northern states, although it should be taken in regions where needed to decrease insect overwintering. Yield of the late fall cutting is generally low, and removal of this forage will increase winterkill and decrease first cutting yield the next spring.

High yield and high quality

For harvest schedules to provide the highest yield of high quality forage, the first two cuttings must be timely. During this time forage quality changes most rapidly and short delays mean low quality forage (Figure 23). Take the first cutting at bud stage or between May 15 and 25 in Minnesota and Wisconsin, and earlier farther south. Take the second cutting 28 to 33 days after the first cut or midbud, whichever

Figure 22. Cutting schedules for different management goals.

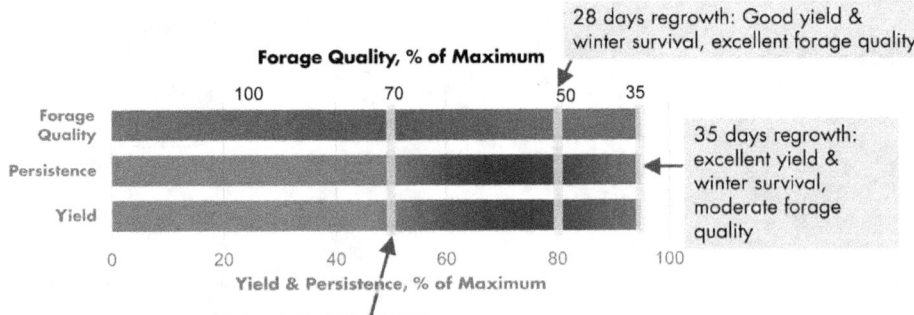

28 days regrowth: Good yield & winter survival, excellent forage quality

Forage Quality, % of Maximum

100   70   50   35

Forage Quality
Persistence
Yield

35 days regrowth: excellent yield & winter survival, moderate forage quality

0   20   40   60   80   100

Yield & Persistence, % of Maximum

Note: Winter survival is enhanced by:
1) Letting one harvest of 3 to 4 cuttings or 2 of 6 or more cuttings go to 10% bloom.
2) Leaving late fall growth through winter.

20 days regrowth: reduced yield & poor winter survival, excessive forage quality

is earlier, and take subsequent cuttings at 38- to 55-day intervals or at 10 to 25% bloom. An early first harvest followed by a short cutting interval gives a high yield of quality forage (Figure 22) while letting one cutting mature to early flower will increase root reserves and stand persistence. The forage quality of alfalfa does not change as rapidly in later cuttings as in earlier cuttings so later cuttings maintain quality to later maturity stages (Figure 23). This slower quality change allows a harvest window of 7 to 10 days. Additional cuttings may be taken if time permits before the required 6- to 8-week rest period prior to the first killing frost. In northern regions, delaying the third cut often results in alfalfa flowering during the 6 weeks before the first killing frost (between September 1 and October 15 in northern states). To prevent loss of persistence, delay harvest until mid- to late October, regardless of the stage of maturity. However, this late-fall cutting will shorten stand life and decrease yield the next spring, so should be cut high (at 6 inches) or not harvested if adequate forage is available. Minnesota researchers found that highest yields came from three cuttings during the growing season with a late-fall cutting. Using this cutting schedule, the percentage of total yield cut at "prime standard" (>150 RFV index) ranged from 32 to 75%.

Fall management

Fall management of alfalfa involves assessing the risk of winter injury and the need for additional forage. The risk of winter injury to alfalfa depends on uncontrollable environmental factors (snow cover, temperature, and soil moisture) and controllable factors (variety, soil fertility, seasonal cutting strategy, stand age, and cutting height).

Uncontrollable environmental factors

- Extended periods of cool temperatures are required in the fall for alfalfa to develop resistance to cold temperatures. Sudden changes from warm to cold reduce hardening.

- A snow cover of 6 inches or more protects alfalfa plants from severe cold. During winters without snow cover, soil temperatures can fall below 15°F, injuring or killing plants.

- Even hardy varieties can be injured or killed by 2 weeks or more of temperatures below 5° to 15°F.

- Warm fall weather (40°F or higher) and midwinter thaws cause alfalfa to break dormancy and have less resistance to freezing.

- Excessively moist soil in the fall reduces hardening and predisposes alfalfa to winter injury. Excess surface and soil moisture can lead to the formation of ice sheets. Ice sheets smother plants by freezing the soil before the plant has hardened. Also, high concentrations of toxic substances—such as carbon dioxide, ethanol, and methanol—accumulate beneath the ice. Ice sheeting frequently occurs in conjunction with midwinter thawing and is more prevalent in poorly drained soils.

Figure 23. Dry matter yields increase with longer intervals between cuttings while forage quality rapidly declines, particularly during first and second cuttings.

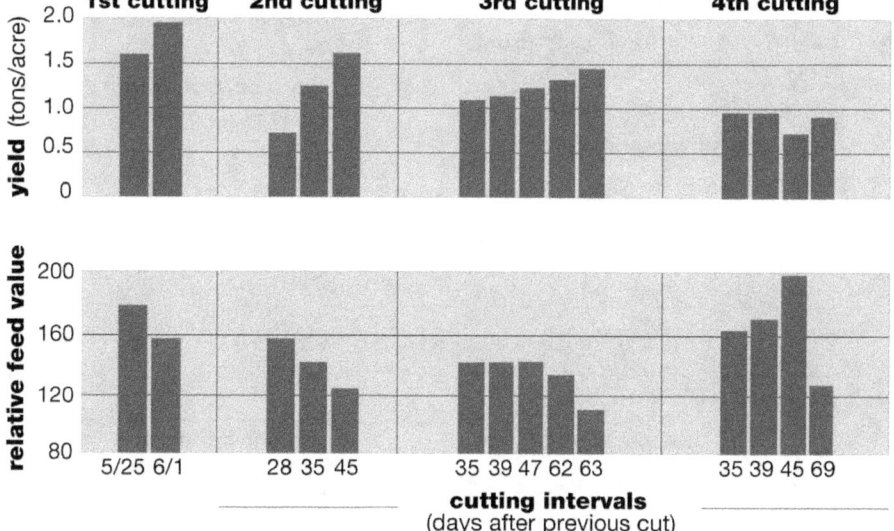

Source: Adapted from Brink and Marten, University of Minnesota, 1989.

Controllable factors

- Select alfalfa varieties with good winterhardiness and moderate resistance to several diseases. These varieties will better tolerate late-fall cuttings.

- Soil fertility management is vitally important for maintaining productive alfalfa stands. Potassium (potash) is particularly important for developing plants that have good winter survival.

- Greater harvest frequency and stand age at harvest increases the potential for winter injury when fall cuttings are taken. When the interval between previous cuttings has been 35 days or less, avoid harvesting during the critical fall period 6 weeks before the first killing frost (between September 1 and October 15 in northern states, later in southern states). This allows plants to enter winter with higher root carbohydrates (Figure 21).

- Young alfalfa stands survive winters better than older stands due to lower disease infestation and less physical damage.

- Stem and leaf stubble remaining in the late fall catch snow and insulate the soil. Alfalfa harvested in October should have a 6-inch stubble left and uncut strips to reduce risk of winter damage.

Winter injury risk

If you score:	Your risk is:
3–7	low/below average
8–12	moderate/average
13–17	high/above average
>17	very high/dangerous

Table 11. Calculate your risk of alfalfa winter injury. Enter the score for answers that describe your situation.

	points	score
1. What is your stand age?		
>3 years	4	
2–3 years	2	
≤1 year	1	
2. Describe your alfalfa variety		
a. What is the winderhardiness?		
Higher than recommended for region	3	
Recommended for region	2	
Lower than recommended for region	1	
a. total		
b. What is the resistance to important diseases in your region?		
No resistance	4	
Moderate or low resistance	3	
High level of resistance	2	
b. total		
Alfalfa variety total score (multiply a and b)		
3. What is your soil pH?		
6.0	4	
6.1–6.5	2	
≥6.6	0	
4. What is your soil exchangeable K level?		
Low (≤ 80 ppm	4	
Medium (81–120 ppm)	3	
Optimum (121–160 ppm)	1	
High (≥161 ppm)	0	
5. What is your soil drainage?		
Poor (somewhat poorly drained)	3	
Medium (well to moderately well drained)	2	
Excellent (sandy soils)	1	
6. What is your soil moisture during fall/winter?		
Medium to dry	0	
Wet	5	

7. Describe your harvest frequency:

Cut interval	Last cutting†		
<30 days	Sept. 1–Oct. 15	5	
	After Oct. 15	4	
	Before Sept. 1	3	
30–35 days	Sept. 1–Oct. 15	4	
	After Oct. 15	2	
	Before Sept. 1	0	
>35 days	Sept. 1–Oct. 15	2	
	After Oct. 15	0	
	Before Sept. 1	0	

	points	score
8. For a mid-September or late October cut, do you leave more than 6 inches of stubble?		
No	1	
Yes	0	
Determine your total score (sum of points from Questions 1–8)		

† Dates listed are for northernmost states; states south of that area should use later dates.
Source: Adapted from C.C. Sheaffer, University of Minnesota, 1990.

Making the decision to cut in the fall requires using the above factors to estimate the risk of winter injury to alfalfa and weighing it against the need for forage. The questions in Table 11 will help you assess the risk of winter injury.

Harvesting the late-fall cutting will increase tonnage for the season and may be more profitable in areas where risk potential is low (see Table 11) and good snow cover is likely and in areas with less severe winters. Minnesota research shows that taking a fourth cutting after October 15 is more profitable than three cuts by September 1 (6 weeks before killing frost) or four cuts by September 15 with no fall cutting for a 4-year-old alfalfa stand. In five-cut systems, the first cutting yield the next spring was lowered by approximately the same amount as the yield from the fall cutting. Root rot was increased and, therefore, stand life was also shortened.

Hay and silage management

Hay-making and silage-making differ in how the moisture content of alfalfa is employed as a strategy in preservation. Fresh alfalfa contains about 80% moisture. Soluble sugars and proteins are dissolved in the forage liquid. When concentrated through wilting, this "juice" provides an ideal medium for the growth of yeasts, molds, and bacteria and for rapid activity of plant enzymes. Appropriate bacterial growth can result in fermentation that produces lactic acid and preserves the material as silage. When forage is dried to hay before harvest, water in the forage evaporates, resulting in a higher concentration of nutrients in the remaining liquid where cell growth and enzyme activity are restricted.

Losses

Each step in the preservation process— mowing, raking, chopping, baling, storing, and unloading—causes a loss of forage dry matter (Figure 24). Some losses result from mechanical action; others are biological processes. Total losses from cutting to feeding are 20% to 30% of the standing crop dry matter in typical hay and silage systems. In hay-making, most of the losses result from mechanical handling and weather damage in the field. In silage-making, most losses occur during storage and feed out.

Quality changes

Most of the dry matter lost from forage during harvest and storage has high nutritional value. More leaves than stems are lost during hay-making, and most protein- and energy-rich nutrients are concentrated in the leaves. Biological processes in silage-making use the most readily available nutrients, such as plant sugars. Thus, in both hay and silage systems, the changes that occur are often detrimental to the quality of the final product.

Minimizing losses

Dry matter losses and quality changes cannot be eliminated in hay preservation, but they can be minimized by using good management practices. The practices for good hay-making are summarized in Table 12.

Figure 24. Dry matter losses during harvest and storage relative to forage moisture content at harvest.

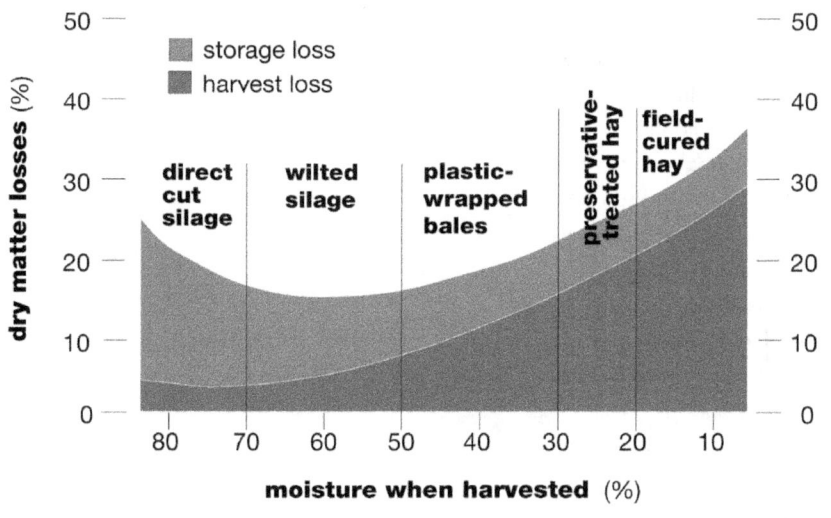

Quality losses during hay-making

- Respiration uses plant sugars, a process that increases NDF and decreases digestibility.

- Rain on hay before baling leaches soluble nutrients (protein and carbohydrates). NDF and ADF increase; digestibility and crude protein decrease. Additional quality is lost from leaf shattering.

- Rainy weather causes delays in harvest. NDF and ADF increase; digestibility and crude protein decrease.

Good hay preservation depends primarily on handling and harvest management. The drying rate, mechanical handling of the forage, and the moisture content at baling all affect the quality of the hay. With proper management, little or no deterioration takes place in the hay during storage.

Quality losses during silage-making

- Respiration uses plant sugars, a process that increases NDF; decreases digestibility and dry matter intake by animals.

- Loss of leaves decreases crude protein.

- Soluble protein can increase in silage during fermentation. Animals on high-performance diets (dairy or growing beef) need insoluble protein, so performance is lowered.

- Acid detergent fiber crude protein is protein made insoluble through the heating during fermentation. Up to 14% is beneficial; more than 14% reduces protein availability to the animal.

Unavoidable losses include those due to field losses, plant respiration, and primary fermentation. Avoidable losses occur from effluent, anaerobic fermentation, and aerobic deterioration in storage structure. Estimates of unavoidable dry matter losses range from 8% to 30%; avoidable losses range from 2% to 40% or higher. The importance of quickly achieving and maintaining oxygen-free conditions has led to improved equipment and techniques for precision chopping, better compaction, rapid filling, and complete sealing.

Alfalfa is more difficult than corn to ferment properly because alfalfa contains fewer soluble carbohydrates relative to protein. For an outline of good silage management practices see Table 13 on the next page.

Table 12. Summary of good hay-making practices.

Practice	Reason	Benefit
mow forage early in day	allow full day's drying	faster drop in moisture less respiration loss less likelihood of rain damage
form into wide swath	increase drying rate	faster drop in moisture less respiration loss less likelihood of rain damage higher quantity and quality
rake at 40–50% moisture content	increase drying rate	faster drop in moisture less respiration loss less likelihood of rain damage less leaf shatter higher quantity and quality
bale hay at 18–20% moisture content	optimize preservation	less leaf shatter inhibits molds and browning low chance of fire higher quantity and quality
store hay under cover	protect from rain, sun	inhibits molds and browning less loss from rain damage higher quantity and quality

Source: Pitt, Cornell University, 1991.

Wide swaths (left) dry faster and have less yield and quality loss than narrow swaths (right).

Table 13. Summary of good alfalfa silage practices.

Practice	Reason	Benefit
minimize drying time	reduce respiration	reduced nutrient and energy losses more sugar for fermentation lower silage pH
chop at correct TLC † fill silo quickly enhance compaction seal silo carefully	minimize exposure to oxygen	reduced nutrient and energy losses more sugar for fermentation reduced silo temperature less heat damage (browning) faster pH decline better aerobic stability less chance of listeria less protein solubilization
ensile at 30–50% dry matter content	optimize fermentation	reduced nutrient and energy losses proper silo temperatures less heat damage (browning) control clostridia prevent effluent flow
bale hay at 1820% moisture content	allow complete fermentation	lower silage pH more fermentation acids better aerobic stability less chance of listeria
unload 2–6 inches/day keep surface smooth	stay ahead of spoilage	limit aerovic deterioration
discard deteriorated silage	avoid animal health problems	prevent toxic poisoning, mycotic infections prevent listeriosis, clostridial toxins

† TLC = theoretical length of cut. Chop alfalfa silage at 3/8-inch TLC.
Source: Pitt, R.E., Cornell University, 1990.

Feeding considerations of hay and haylage

A widely used rule of thumb in formulating rations for lactating dairy cattle is that one-third of the diet be forage, one-third concentrate, and the remaining one-third either forage or grain, depending upon the quality of the forage fed. As long as the NDF stays below 1.2% of body weight (so ration intake is not reduced) and 41 or 42% Nonfiberous carbohydrate (NFC). By feeding high quality alfalfa in place of lower quality forages, dairy producers can decrease the amount of concentrates that must be fed and can increase the utilization of forage. The lactation study in Figure 25 shows concentrates cannot supply the energy required at high production levels when the quality of the forage is too low.

How alfalfa is harvested and preserved has been the focus of many research studies, but no clear advantage in animal performance has been demonstrated for harvesting and storing alfalfa either as hay or haylage. Harvesting alfalfa at higher moisture contents will decrease field losses but will increase storage losses unless forage is kept in airtight silos or silage tubes.

Figure 25. Fat-corrected milk (FCM) yield as influenced by change in alfalfa maturity and concentrate level.

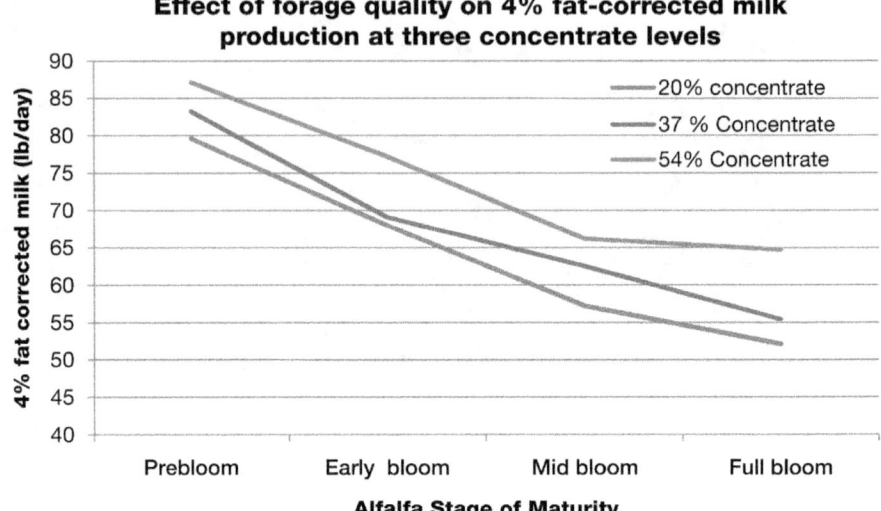

Advanced Techniques

Drying agents, preservatives, and silage inoculants

To speed drying, use a drying agent in addition to mechanical conditioning. These products, either sodium or potassium carbonate, should be applied to alfalfa as it is cut. They will shorten drying time by 5 to 24 hours. Drying agents do not work on grasses. These products cost $2 to $6 per acre and require large volumes of water for application.

Preservatives allow hay to be baled at higher moisture contents than can normally be stored: above 14% for bales larger than $3\frac{1}{2}' \times 3\frac{1}{2}'$; 16% for $2\frac{1}{2}'$ to $3\frac{1}{2}'$ square bales or round bales; and 20% for small square bales. These products are only cost effective if their use prevents rain damage, so apply only when rain is imminent. Propionic acid is the most effective chemical preservative. Ammonium propionate is less caustic than propionic acid and equally as effective per unit of propionate. Acetic acid is only half as effective as propionic acid as a preservative. In all cases the amount needed for preservation is in relation to the moisture content of the hay (Figure 26).

Silage inoculants provide the lactic acid-forming bacteria required for good haylage or silage fermentation. These products (either microbial or enzyme formulations) are beneficial when naturally occurring populations of lactic acid-forming bacteria are low and plant carbohydrate levels high. In the northern United States, these conditions occur on all early- and late-season cuttings when the drying time has been less than 2 days (Figure 27). Bacterial inoculants must be stored in cool places and contain 10^6 Lacto-bacillus plantarum colony forming units (cfu) per gram. To be effective, the inoculant must be uniformly mixed throughout the forage. A liquid applicator on the chopper or on the blower is the preferred method of application.

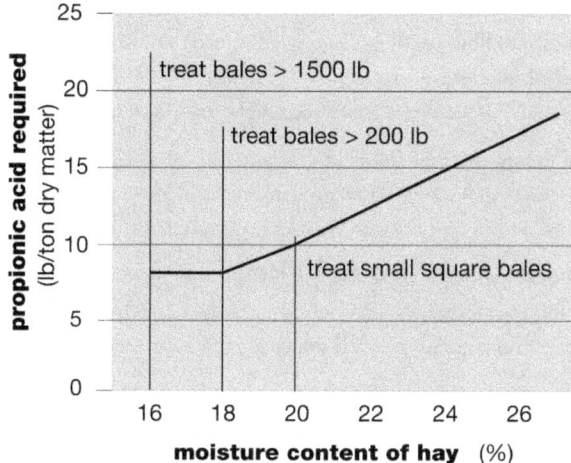

Fig. 26. Propionic acid needed to preserve hay.

Source: Undersander, University of Wisconsin, 1999.

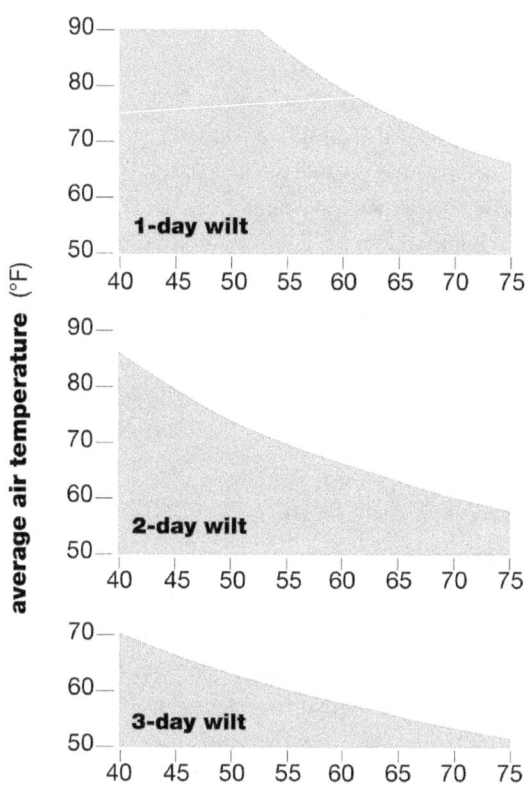

Figure 27. Conditions for profitable use of inoculant on silage. Shaded areas indicate profitable conditions.

Source: Adapted from Muck, USDA, 1993.

Forage quality terms

Acid detergent fiber (ADF) is the percentage of highly indigestible and slowly digestible material in a feed or forage. This fraction includes cellulose, lignin, pectin, and ash. Lower ADF indicates a more digestible forage and is more desirable.

Neutral detergent fiber (NDF) is the percentage of cell walls or fiber (i.e., cellulose, hemicellulose and lignin with some ash), not including pectin. NDF is inversely related to animal intake and the energy that an animal can derive from a feedstuff.

Neutral detergent fiber digestibility (NDFD) is the percentage of the NDF that is digested by animals in a specified time period (usually 24, 30, or 48 hours). NDFD is positively related to animal intake and the energy that an animal can derive from a feedstuff.

Total digestible nutrients (TDN) is the sum of digestible crude protein, nonfibrous carbohydrate, fat (multiplied by 2.25), and digestible NDF minus 7.

Relative forage quality (RFQ) is an index used to rank forages by potential intake of digestible matter where 150 and above is considered milking dairy quality feed and lower indices are needed for other categories of animals (Figure 19).

Relative Forage Quality Calculations for Legumes

1. Calculate digestible dry matter of forage (% of Dry matter)
   $$TDN = (NFC \times .98) + (CP \times .93) + (FA \times .97 \times 2.25) + (NDFn \times (NDFD/100) - 7$$
   where:
   CP = crude protein (% of DM)
   EE = ether extract (% of DM)
   FA = fatty acids (% of DM) = ether extract − 1
   NDF = neutral detergent fiber (% of DM)
   NDFCP = neutral detergent fiber crude protein
   NDFn = nitrogen free NDF = NDF − NDFCP, else estimated as NDFn = NDF × .93
   NDFD = 48-hour in vitro NDF digestibility (% of NDF)
   NFC = non fibrous carbohydrate (% of DM) = 100 − (NDFn + CP + EE + ash)

2. Calculate dry matter intake of forage (% of body weight)
   $$DMI = 120/NDF + (NDFD - \text{laboratory average digestibility for alfalfa}) \times .374/1350 \times 100$$

3. Calculate Relative Forage Quality
   $$RFQ = (DMI, \% \text{ of BW}) \times (TDN, \% \text{ of DM})/1.23$$

Crude protein (CP) is a mixture of true protein and nonprotein nitrogen. It is determined by measuring total nitrogen and multiplying this number by 6.25. Crude protein content indicates the capacity of the feed to meet an animal's protein needs. Generally, moderate to high CP is desirable since this reduces the need for supplemental protein. Forage cut early or with a high percentage of leaves has a high CP content.

Rumen undegraded protein (also called bypass protein) is that portion of the protein not degraded in the rumen. Some bypass protein is needed for high producing dairy animals.

www.ingramcontent.com/pod-product-compliance
Lightning Source LLC
LaVergne TN
LVHW081201190326
834317LV00012B/413